# Data Science for Agricultural Innovation and Productivity

Edited by

**S. Gowrishankar**
*Department of Computer Science and Engineering*
*Dr. Ambedkar Institute of Technology*
*Bengaluru, Karnataka 560056*
*India*

**Hamidah Ibrahim**
*Department of Computer Science*
*Faculty of Computer Science and Information Technology,*
*Universiti Putra*
*43400, Selangor, Malaysia*

**A. Veena**
*Department of Computer Science and Engineering*
*Dr. Ambedkar Institute of Technology*
*Bengaluru, Karnataka 560056*
*India*

**K.P. Asha Rani**
*Department of Computer Science and Engineering*
*Dr. Ambedkar Institute of Technology*
*Bengaluru, Karnataka 560056*
*India*

&

**A.H. Srinivasa**
*Department of Computer Science and Engineering*
*Dr. Ambedkar Institute of Technology*
*Bengaluru, Karnataka 560056*
*India*

**Data Science for Agricultural Innovation and Productivity**

Editors: S. Gowrishankar, Hamidah Ibrahim, A. Veena, K. P. Asha Rani & A. H. Srinivasa

ISBN (Online): 978-981-5196-17-7

ISBN (Print):  978-981-5196-18-4

ISBN (Paperback): 978-981-5196-19-1

need for a court order if at any point you breach any terms of this License Agreement. In no event will any delay or failure by Bentham Science Publishers in enforcing your compliance with this License Agreement constitute a waiver of any of its rights.

3. You acknowledge that you have read this License Agreement, and agree to be bound by its terms and conditions. To the extent that any other terms and conditions presented on any website of Bentham Science Publishers conflict with, or are inconsistent with, the terms and conditions set out in this License Agreement, you acknowledge that the terms and conditions set out in this License Agreement shall prevail.

**Bentham Science Publishers Pte. Ltd.**
80 Robinson Road #02-00
Singapore 068898
Singapore
Email: subscriptions@benthamscience.net

**BENTHAM SCIENCE**

# CONTENTS

# FOREWORD

Agriculture, which contributes 15.4% to the GDP of India, was the foundation on which we built our nation and this has undergone several development phases. Agriculture is the practice of growing the food and cash crops that humans need, and its techniques and technology have advanced quickly. The selection of farming methods depends on climate, terrain, traditions, and the financial condition of the farmer. Depending on the exposure to technology, the availability of land, capital, and skilled labor, the aforementioned techniques can be used in large-scale or small-scale farming.

Each crop needs different growth conditions, differing harvesting seasons, and distinct attention. It is hard to deny the common techniques of agriculture that farmers across the world use. These include ploughing, the use of modern machinery, fertilizers, various types of seeds, *etc.* But looming on the horizon is India's population boom, which is set to rise to 1.515 billion by 2030 from 1.417 billion in 2022. This has increased pressure on the agriculture sector.

The full potential of agricultural innovation has to be fully realised in many emerging nations. Understanding and implementing the innovation drivers and procedures that are essential for realising innovation's potential and sparking it are the keys to success. Unfortunately, the cost of the agricultural cycle, beginning from the preparation of soil to the selling of the produce, has ratcheted up, along with the risks. As this book shows, relevancy in information is key and keeping track of crops, the environment and the market may help farmers to make better decisions and ease problems related to agriculture. Technologies like IoT, Machine learning, deep learning, hydroponics, and smart farming can attain information and process it. The authors of this book present a powerful argument for the application and implementation of such aforementioned technologies that will surely help alleviate many challenges we currently face, if not outright prevent it.

**Sharath Malve**
Technology Expert
Hewlett Packard Enterprise
Bengaluru, Karnataka, India

# PREFACE

Agriculture can provide for one of humanity's essential needs, which is food. Around the globe, agriculture is a source of employment besides providing for humankind's basic requirements. Agriculture now entails significantly more than merely planting a seed, raising a cow, or capturing a fish. To feed a huge population, an entire environment and a spate of individuals must work together. Innovation enables us to achieve more and better with less. Innovation is driven not just by technology breakthroughs, but also by creative methods of organising farmers and linking them to the information they want. Many smallholder farmers throughout the world continue to cultivate in the same manner that their forefathers did hundreds of years ago. Traditional farming methods may continue to be effective for some, but new tactics may assist many in significantly improving yields, soil quality, and natural capital, as well as food and nutrition security. Farmers can be grouped in novel ways to ensure that information reaches them more readily and efficiently. The kind and style of the extension itself have altered significantly throughout time. For example, developments in satellite mapping and information and communication technologies (ICTs) are already altering more traditional agricultural extension activities. Farming is getting more accurate and productive as a result.

The number of farm records in electronic format is growing daily, and informatics and data analysis are essential to analyse these huge records with diverse datasets.

This book reveals the use of different sensors to collect diverse farm data, which comprise chemical/pesticide tracking, harvest and yield records, planting records, shipping records, labor tracking, weather data, *etc.* Sustainable Agriculture focuses on meeting current requirements without compromising the ability of future generations to meet their own. An agriculture system that enables farms of all sizes to be profitable and contribute to their local economies is one that is both economically and socially sustainable. A system like this would prioritise people and communities over corporate interests, support the next generation of farmers, provide everyone with access to healthy food, and support the next generation of farmers. Because of the widely varying scales, dimensions, and volumes of electronic farm data, multidisciplinary solutions are needed for visual depiction and digital characterisation. Gains in agriculture have improved thanks to developments in machine learning. By offering detailed advice and insights about the crops, machine learning is a current technology that helps farmers reduce farming losses.

Farmers, vendors, theoreticians and engineers have used various software applications and servers for these uses. Because of problems with disease identification, a lack of interoperability brought on by vendor-locked agriculture systems, and security/privacy concerns regarding data storage, sharing, and usage, the agriculture domain is well known for suffering from heterogeneous and uneven data, delayed farm communications, and disparate work flow tools.

The material in the book is presented in a way to encourage researchers to think and indicate concepts that are introduced, which can solve real-world problems in the agricultural domain. Research and extension are crucial to innovation pathways. The contents of this book focus on,

Smart Farming - The future of agricultural technology is big data collection and analysis in agriculture to improve operational efficiency and reduce labour expenses. Based on a more precise and resource-efficient strategy, smart farming has the potential to offer more productive and sustainable agricultural output. IoT has encouraged the assumption that a smart network of sensors, actuators, cameras, robots, drones, and other connected devices would provide agriculture with new levels of control and automated decision-making, allowing for a lasting ecosystem of innovation.

Artificial Intelligence - It is progressively growing as a component of the agricultural industry's technological growth. It is playing a critical role in the agriculture sector and is altering the industry. AI protects the agriculture sector against a variety of threats, including climate change, population expansion, labour shortages, and food safety. The goal is to boost global food production and will not only help farmers improve efficiency, but they will also increase crop quantity and quality and ensure crops reach the market faster.

Machine Learning - With the emergence of IoT and other technologies, a fertile ground for real-time monitoring has emerged. Machine learning solutions in agriculture rely on real-time data to provide farmers with exponential advantages. AI and machine learning are powerful catalysts for improving remote facility security, yields, and pesticide efficacy.

Deep Learning- It is a relatively new, innovative approach for image processing and data analysis, with promising results and enormous potential. Deep learning has recently entered the agricultural sector after being effectively employed in other disciplines.

Hydroponics system - It is the cultivation of plants in a controlled setting. While indoor farming is not a new phenomenon, hydroponic farming, a more recent discovery, simplifies the growing process even more by eliminating all unneeded components of traditional farming. Small farmers, amateurs, and business companies all employ hydroponic production systems.

Robotics in Agriculture - Robotics will undoubtedly bring about an agricultural revolution. Although the road ahead is not very smooth, we must assess the feasibility, sustainability, and efficiency of providing the world's food demands. However, it will be exciting to observe how farmers, agribusinessmen, and consumers will use the potential of robotics and digital-mechanization to define the future of this sector.

Internet of Green Things - Green IoT is an evolution of IoT that reduces emissions and pollution through many elements while also having low operational costs and power usage. Green IoT is the future, especially as the world seeks new methods to combat climate change; this new domain offers several chances for enterprises.

Crop health monitoring - It encompasses the monitoring of several factors, such as temperature, humidity, precipitation, insect intrusion, and seed and soil quality, allowing for improved crop quality and health decisions. This requires rapid involvement of farm managers in cases of emergency, such as overnight freezing or pest incursions, even from remote areas where they are not accessible on-site.

Application of sensors in agriculture - Given the current circumstances and their adverse influence on traditional farming techniques, agriculture must be carried out more wisely, utilising innovative and cutting-edge technologies. It is the only method to give a solution and fulfil the world's population's infinite and expanding requirements. Farmers may now remotely record their crops and monitor their efficacy, manage agricultural pests, and take quick action to safeguard their crops from environmental threats by using smart sensors in agriculture.

Climate Adaptation Strategies - Climate change, which primarily affects hydro-meteorological threats, is a fact that is influencing the planet in a variety of ways. It manifests itself in a variety of ways, including a rise in the frequency and severity of floods, droughts, and high temperatures. Climate change has caused droughts, other extreme weather occurrences, and meteorological disasters in many nations in recent years. Effective management of climate change-induced difficulties causes localised techniques that may differ from one region of the world to the next, and even within a single country.

Web 3.0 for Farming- Web3, the third generation of the internet and simple video-based technologies in local languages, has the potential to transform agriculture. It refers to initiatives taken to develop a decentralised form of the internet based on blockchain technology and on user ownership, which has the potential to reverse old data paradigms and return power to farmers.

The precise focus of this handbook will be on the potential applications and use of data informatics in the area of the agriculture domain.

**S. Gowrishankar**
Department of Computer Science and Engineering
Dr. Ambedkar Institute of Technology
Bengaluru, Karnataka 560056
India

**Hamidah Ibrahim**
Department of Computer Science
Faculty of Computer Science and Information Technology
Universiti Putra
43400, Selangor, Malaysia

**A. Veena**
Department of Computer Science and Engineering
Dr. Ambedkar Institute of Technology
Bengaluru, Karnataka 560056
India

**K. P. Asha Rani**
Department of Computer Science and Engineering
Dr. Ambedkar Institute of Technology
Bengaluru, Karnataka 560056
India

&

**A. H. Srinivasa**
Department of Computer Science and Engineering
Dr. Ambedkar Institute of Technology
Bengaluru, Karnataka 560056
India

# List of Contributors

| | |
|---|---|
| **Ahmad Muhammad Makarfi** | Department of Agricultural Economics and Extension, BUK, Kano, Nigeria |
| **Bogala Mallikharjuna Reddy** | ACPL, Technology Business Incubator, University of Madras, Guindy Campus, Chennai, India |
| **Carmel Sobia M.** | Department of Electrical and Electronics Engineering, PSR Engineering College, Sivakasi, Tamilnadu, India |
| **Chetan Khadse** | National Institute of Technology, Hamirpur (H.P.), Himachal Pradesh 177005, India |
| **Deepak S. Sakkari** | Networking and IoT lab, Presidency University, Sri Krishna Institute of Technology, Bangalore, India |
| **Galiveeti Poornima** | Networking and IoT lab, Presidency University, Sri Krishna Institute of Technology, Bangalore, India |
| **Gopal Rawat** | National Institute of Technology, Hamirpur (H.P.), Himachal Pradesh 177005, India |
| **Jayalakshmi Murugan** | Department of Computer Science and Engineering, School of Computing, Kalasalingam Academy of Research and Education, Krishnankoil, Tamilnadu, India |
| **Maharajan Kaliyanandi** | Department of Computer Science and Engineering, School of Computing, Kalasalingam Academy of Research and Education, Krishnankoil, Tamilnadu, India |
| **Nakatha Arun Kumar** | BMS College of Engineering, Bangalore, India |
| **Paritala Venkateswara Rao** | School of Computer Science and Engineering, Presidency University, Bangalore, India |
| **Prasenjit Pal** | Department of Fisheries Extension, Economics and Statistics, College of Fisheries, CAU (I), Lembucherra, Tripura, India |
| **Sukruth Gowda M.A.** | Networking and IoT lab, Presidency University, Sri Krishna Institute of Technology, Bangalore, India |
| **Sathish S. Kumar** | RNS Institute of Technology, Bangalore, India |
| **Supriya Jaiswal** | National Institute of Technology, Hamirpur (H.P.), Himachal Pradesh 177005, India |
| **Sohit Sharma** | National Institute of Technology, Hamirpur (H.P.), Himachal Pradesh 177005, India |
| **Sadiq Mohammed Sanusi** | Department Of Agricultural Economics & Extension, Federal University Dutse, P.M.B. 7156, Dutse, India |
| **Singh Invinder Paul** | Department of Agricultural Economics, SKRAU, Bikaner, India |
| **Sandeep Poddar** | Lincoln University College, Petaling Jaya, Selangor Darul Ehsan, Malaysia |
| **R. Sapna** | Department of Information Technology, Manipal Institute of Technology, Bangalore, Manipal Academy for Higher Education, Manipal, India |

**Ravva Akash Guptha**       School of Computer Science and Engineering, Presidency University, Bangalore, India

**Rwitabrata Mallick**       Department of Environmental Science, Amity School of Life Science, Amity University, Madhya Pradesh, India

**Raavi Sai Pranay**       School of Computer Science and Engineering, Presidency University, Bangalore, India

# CHAPTER 1

# Digital Twin for Smart Farming

**Galiveeti Poornima**[1,*], **Deepak S. Sakkari**[1] and **Sukruth Gowda M.A.**[1]

[1] *Networking and IoT lab, Presidency University, Sri Krishna Institute of Technology, Bangalore, India*

**Abstract:** One of the disruptive technologies that will emerge in the 21st century is the digital twin, which is a digital copy of any physical object that may exist in any setting. Many industries heavily rely on digital twin technology to produce high-quality products that can be shipped throughout the world with no loss in efficiency. The initial efforts have been made by the agricultural sector toward the implementation of digital twin technology in farming and other types of activities. It has already begun to apply vertical farming together with other crucial cutting-edge technologies in a chosen number of smart cities.

**Keywords:** Disruptive technologies, Efficiency, Digital twin, Smart cities.

## INTRODUCTION

Without trustworthy and up-to-date information regarding farm operations, modern agricultural production is not viable. The use of digital technology in agriculture, such as sensing and monitoring devices, sophisticated analytics, and smart equipment, is becoming more necessary. Fast-evolving technologies like the cloud, the Internet of Things, big data, machine learning, augmented reality, and robots are driving a shift in agriculture toward "smart farming" systems [1-4]. One way to look at smart farming is as the natural progression of precision agriculture [5]. In smart farming, management activities are not just dependent on accurate location data but also on context data, situational awareness and event triggers. One way to think about a smart farming system is a cyber-physical control cycle that integrates sensing and monitoring, intelligent analyses and planning, and intelligent control of farm operations for all relevant farm processes (also known as a "whole farm management perspective").

Using digital information that is (near) real-time rather than on-site direct observation and physical labour, farmers may remotely monitor and manage

---

* **Corresponding author Galiveeti Poornima:** Networking and IoT lab, Presidency University, Sri Krishna Institute of Technology, Bangalore, India; E-mail: galiveetipoornima@presidencyuniversity.in

**S. Gowrishankar, Hamidah Ibrahim, A. Veena, K. P. Asha Rani & A. H. Srinivasa (Eds.)**

operarations in smart farming systems. Therefore, farmers  are immediately notified of any issues or impending problems. One may examine a high-quality digital picture of the plant, animal or machine in question from the comfort of their workstation or smartphone to see how things are doing out in the field or stable. Simultaneously, machine learning algorithms enhance the digital view by adding object-specific evaluations and recommendations. Farmers are able to replicate both remedial and preventative activities and assess the effect of such simulations on the digital depiction. Last but not least, the selected intervention may be carried out remotely, and the farmer can utilize the digital view once again to determine whether or not the issue has been resolved as anticipated. As this smart farm management cycle grows more autonomous, the farmer will no longer need to intervene manually, which is another thing that may be anticipated. It's fair to state that everything on the farm (crop, field, cow, equipment) is gradually being virtualized and, therefore, increasingly, remote-controllable. The concept of a digital twin is an interesting metaphor that can be used to describe this development.

Even though there are different ways to define a "Digital Twin", which will be covered later, in general, a "Digital Twin" is a digital copy of a real-world object that acts and changes in the same way in a virtual space [6, 7]. Using Digital Twins as a core tool for farm management decouples physical flows from their planning and control. A digital twin eliminates important limitations related to space, time, and human observation. The need for farmers to be physically close to their crops would be eliminated, which would make it possible to execute, monitor, manage and coordinate farming tasks remotely *via* automation. This makes it possible to decouple the physical flows of agricultural activities from the informational components of those processes. Data from sensors and satellites, for example, may provide context to a digital twin that would otherwise be impossible to get directly from the physical object.

## TECHNOLOGIES USED IN SMART FARMING

While innovation and digital transformation are occurring across many sectors, they are particularly crucial for the future of the planet and the well-being of humans in agriculture. The World Economic Forum projects that the world population will reach 9.8 billion by 2050, which implies that we may need to produce twice as much food as we do now without considerably straining natural resources like land and water.

However, there are good reasons to maintain a hopeful outlook. Throughout history, however, inventive people have found ways to overcome this problem. 8,000 years ago, during the first agricultural revolution, the plough revolutionized production. In the 1800s, developments such as the seed drill introduced a degree of mechanization to farming. The middle of the 20ᵗʰ century saw a number of significant advancements made in the fields of artificial fertilizer and plant science.

At this point in time, we have entered the fourth era of agriculture. The rate of innovation is over the roof, and venture capital funding is flooding in. During 2020, that is, during COVID-19, Finistere Ventures and PitchBook Data projected a 22.3% increase in funding for this sector in 2020, reaching a total of $22.3 billion. In order to put this into perspective, the sum amount of investments made since 2010 is now $65.4 billion.

There is clearly longer any room for speculation on the digital transformation of agriculture. It is not a hoax. In addition, it has a significant influence on the agricultural industry as a whole. Here are some instances.

- **Drones that plant rice seeds:**

In April 2020, Chinese drone company XAG demonstrated rice sowing in Guangdong. First, it asked two employees to scatter 5kg of rice seeds by wading through a flooded paddy area. It took an hour and a quarter to complete the arduous task.

Then it deployed its XAG Xplanet drone on the same duty (Fig. **1**). The unmanned aircraft system flew along a path that had been pre-programmed for it and dropped rice seeds from the sky. The task was finished in one minute and one hundred and twenty seconds. Farming using intelligence 1: labour from humans 0.

**Fig. (1).** XAG Xplanet drone (Courtesy: https://www.xa.com/en/xp2020).

It is impossible to understate how much of an influence drones have had on farming. Drones improve productivity while also being safer for workers and the environment. According to Xag, compared to conventional methods, this technology may use up to 90 compared to conventional methods, this technology may use up to 90% less water and 30% less chemicals. It also has a higher degree of precision. Using a technology known as JetSeed, it can discharge seeds at speeds of up to 18 metres per second. This prevents any seeds from being carried away by the wind. The end result is an efficiency that is 80 times greater than that of hand sowing.

The use of commercial drones is one of the areas of the Internet of Things (IoT) that is expanding at the quickest rate, and 5G will play an important part in making this expansion possible.

• **Smart agriculture and the iot: a 'ball' to keep grain fresh:**

A smart sensor has the potential to be a game-changing piece of technology for an industry as scattered and out of the way as farming. It has the potential to significantly cut waste while simultaneously increasing production.

One such device is GrainSage, which is manufactured by Telesense as shown in Fig. (**2**).

It's a sensor that's been inserted in a ball, and the business says it looks like something a dog might use as a chew toy. The gadget is then tossed onto the existing heap of grain by the farmers. The ball is programmed to provide information numerous times each day on the temperature and humidity levels in the building.

**Fig. (2).**  GrainSage by Telesense to keep grain fresh (Courtesy: https://www.futurefarming.com/smart-farming).

The cloud receives this data wirelessly, and TeleSense's machine learning algorithms analyse it there to find significant patterns before transmitting it to the TeleSense app.

And how that LPWAN has been developed, there is an energy-saving, Internet of

Things (IoT)-optimized connection that can link these sensors for up to ten years on a single battery charge.

### • The robot that looks after chickens:

There's nobody else around here except us hens… not to mention a robot dubbed "The ChickenBoy" that is dangling from the ceiling as shown in Fig. (**3**).

**Fig. (3).**  The ChickenBoy analysis robot (Courtesy: https://www.robotsscience.com/).

Start-up company Faromatics, located in Barcelona, has garnered a lot of attention for its EU-funded creation: a robot that enables chicken farmers to independently monitor their flock. The device has also received a number of accolades in the field of smart farming. The rail-mounted robot moves along the ceiling and utilizes a series of sensors to monitor heat feeling, air quality, light and sound in chicken housing.

Using a cloud service, farmers may set up The ChickenBoy to notify their phones through push notifications.

### • 5G Farming: helping salmon farmers to 'see' one million fish:

One of the biggest challenges for smart farming technology so far has been connecting its sensors and tracking devices. Farms are large; they are often located in isolated areas. Therefore, establishing a connection between these things *via* 3G or 4G might be challenging. Meanwhile, only the largest and most resourceful farmers are capable of establishing. Wi-Fi access over thousands of acres of land.

5G farming offers solutions to practically all of these issues (Fig. **4**). Low-latency and accessible in any location, 5G can connect 100 times more devices per square kilometer than 4g. It can transfer data at rates up to one hundred times faster than before.

**Fig. (4).** 5G farming (Courtesy: https://www.cubictelecom.com/blog/5g-agriculture-smart-farming/).

A small number of farmers handle one million fish using a centralized feeding system in this region. They use a number of cameras that are positioned close to the cages to watch the procedure. This does not always have a simple solution. The water may get cloudy at times. Sometimes cameras have problems. During the winter, Norway's days are mostly cloudy and gloomy.

When farmers' vision is impaired, they run the danger of over-or under-feeding their animals, which may lead to pollution and spillage.

The issue might be remedied with higher-resolution cameras. However, the video cannot be sent back to the central site using the current 4G networks' limited capacity. However, problems with fibre optic cables are common when travelling vast distances.

This is the motivation for the 5G agricultural experiment that is being hosted by Telenor and the Aquatech company Bluegrove. 5G is dependable and easily available, and it can handle the bandwidth demands of high-definition real-time video. For the Sinkaberg Hansen Gjerdinga use case, the applications were set up and processed on an edge server. This reduced the delay to a minimum. Bluegrove estimates that Sinkaberg Hansen might save up to 50 million NOK in annual costs as a result of the partnership.

• **Soil health in smart farming: machine learning gets its hands dirty:**

One of the most well-known uses of machine learning is found in the field of medicine. In this application, computer programmes may analyze massive data sets to look for clues that might point ot the presence of illness. The same work may take a human expert years to complete.

It is very clear that the same pattern recognition applies to the technologies behind smart farming. In today's market, a variety of companies provide solutions of this sort. The first example is Pattern Ag. Machine learning is used to identify several plant diseases and pests. For instance, it is able to identify a single rootworm egg in a sample of one pound of soil.

The system is capable of both diagnosing and resolving the issue at hand. It then utilizes that information to determine which herbicides to use and how much of each to use (Fig. **5**). This is one of the numerous agricultural advances that, in comparison to previous methods, is both more effective and less harmful to the environment.

In the United States, rootworm pesticides are used to protect 55% of fields, according to Pattern Ag, even though only 18% are in financial danger from these pests. In the absence of accurate data, farmers may overprotect against rootworms.

Feeding 9.8 billion mouths by 2050 is one of the largest challenges mankind faces. However, human ingenuity has historically been able to overcome obstacles of a similar kind. There is strong cause to be optimistic about the future thanks to the technological advancements that have been made in agriculture.

**Fig. (5).**  Teralytic: Wireless NPK sensor for monitoring soil health (Coutesy: https://www.producer. com/crops/independent-probes-take-the-measure-of-the-soil/).

## DEFINITION OF DIGITAL TWIN

The field of agriculture is one that is not only notoriously difficult but also one that is consistently subject to development. Because they are dependent on natural conditions such as weather, diseases, soil conditions, seasonability, and climate, production processes are basically dynamic [8]. In addition to this, producers are obligated to meet the strict standards set forth by consumers and society with regard to concerns of food security, food safety, sustainability, and health. These requirements can be broken down into four categories: As a result, farms must not only be highly efficient, but also meet stringent quality and environmental standards, as well as react to changing market conditions. In addition to this, it is imperative that farms operate as effectively as they possibly can. Farmers' managerial obligations are significantly put under pressure as a result [8, 9]. Timely observation of agricultural operations necessitates constant reevaluation of production strategies and rescheduling of planned activities.

## DIGITAL TWIN TYPOLOGY

The portion of the product's lifespan shown in Fig. (**6**) and referred to as the "utilization phase" is where the majority of the focus of Digital Counterparts is directed [10]. This is the time when digital twins are linked to their corresponding physical twins in the actual world. In this stage, Digital Twins can be used to maintain track of physical objects, prescribe optimal states for those objects, forecast how those objects will change, and even make remote adjustments to the actual objects' existing conditions. Even before the stage in which they are used, Digital Twins can already be developed to characterize and reproduce the states and behaviors of their real-life twins who have not yet been born. This can be done even before the stage in which they are used. This might be accomplished prior to the stage of consumption. Last but not least, digital duplicates can be used to reconstruct the previous states of physical objects even after the initial period of their utilization has come to an end. Based on Redelinghuys *et al.*, [11], Lepenioti *et al.*, [12], we came up with the following definitions for six distinct types of digital twins:

• **Imaginary Digital Twin**: A mental representation of something that does not yet exist in the actual world; also known as a concept. It defines the information that is necessary to actualize its physical twin, including examples such as functional requirements, 3D product models, material and resource specifications, production models, and disposal and recycling specifications. In addition, it defines the information that is necessary to actualize its digital twin [13]. Imaginary twins have the extra capability of replicating the behavior of designed but as of yet unrealized items within acceptable tolerance limits. This can be done provided that the behavior is consistent with the design [14].

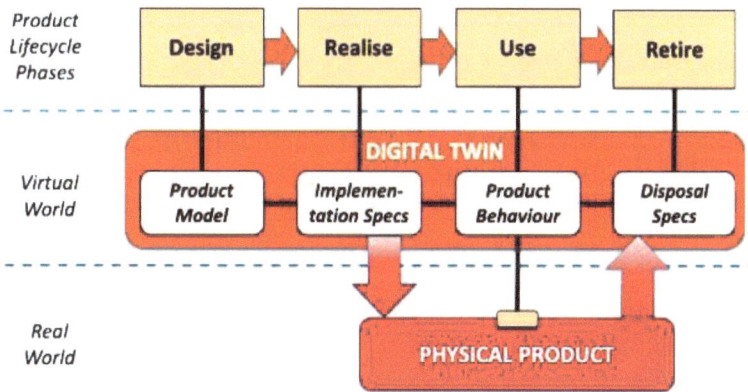

**Fig. (6).** Role of Digital Twins during Product Life Cycle.

• **Monitoring Digital Twin**: A graphical representation of the actual state, behavior, and path that a real-world physical object is currently following. It is connected in (near) real-time to its physical counterpart and is used to monitor its condition as well as its operations and the environment outside of it. In addition to describing what is happening or has happened with the connected physical object, a Digital Twin may also diagnose the underlying causes of these events by drawing connections between the monitored object and its surrounding environment.

• **Predicting Digital Twin**: A computer projection of the future states and behaviors of physical entities, developed using tools from the field of predictive analytics such as statistical forecasting, simulation, and machine learning techniques. Predictions are derived dynamically by using data from the physical twin that is nearly real-time.

• **Prescriptive Digital Twin**: A smart digital item that adds intelligence to real-world things in order to advise remedial and preventative measures, often on the basis of optimization algorithms and expert heuristics. These suggestions are provided by the intelligent digital object as a result of its analysis. By using the results from both predictive and monitoring twins, prescriptive twins can then advise on the best course of action to take. Decisions on the proposed actions are still made by people who are also responsible for triggering remote or on-site execution.

• **Autonomous Digital Twin**: It functions without the need for any involvement from humans, either on-site or remotely, and is completely capable of controlling the behavior of real-world objects. Self-learning, self-diagnosing, and user-preference-adjusting are all within reach for self-driving systems like autonomous twins [15].

• **Recollection Digital Twin**: It captures and stores all of the information about a thing's past existence, even if that thing no longer exists in the present. To put it another way, remembrance twins make up the digital memory of, say, a farm. This type of Digital Twin is frequently disregarded, despite its importance in reducing the environmental impact of disposals and optimizing the next generation of things [16]. In the context of agriculture, recollection twins are of crucial importance for tracing products back to their source in the case that there are problems regarding food safety, as well as for the purpose of complying with regulations for sustainability.

## DIGITAL TWINS IN FARM MANAGEMENT

The agricultural industry is famously challenging and is always undergoing improvement. The weather, illnesses, soil, and seasonality play huge roles in the production process, making it highly dynamic [17]. In addition to this, producers are expected to follow the stringent criteria set forth by consumers and society with regard to concerns of food security, food safety, sustainability, and health. These requirements can be broken in a number of ways. As a result, farms must not only be highly efficient, but also meet demanding quality and environmental standards, as well as adapt to changing market situations. In addition to this, it is of the utmost importance that agricultural operations be carried out in the most efficient manner possible. This considerably increases the pressure on farmers' managerial responsibilities [18]. By facilitating the separation of the physical and data aspects of farm management, digital twins can substantially increase the available control options [19], as shown in Fig. (**7**). Timely observation of agricultural operations requires regular reevaluation of production strategies and rescheduling of planned activities. Despite this, putting Digital Twins to use in agricultural management is a challenging task for at least three distinct reasons [20].

For starters, the highly dynamic agricultural production system (process dynamics) has criteria that exceed those of many other industries in terms of Digital Twin's ability to mimic dynamic behavior [21]. It is very difficult to gain seamless access to object data in an environment that is as dynamic as the one we are discussing since it is difficult to ensure the data's integrity while also honoring use rights, safety, and security. In addition, real-time synchronization might be challenging in rural areas because these locations sometimes have restricted coverage and bandwidth.

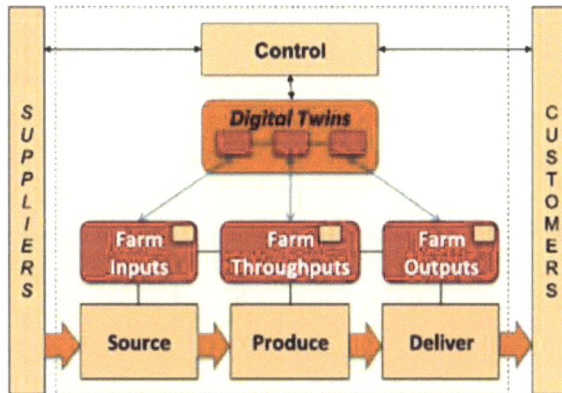

**Fig. (7).** Virtual Control of farming enabled by Digital Twin.

Second, the products of agriculture are living creatures; this means that they are inherently distinct from one another and may be differentiated from one another based on complex patterns of activity. In addition, smart farms consist of a wide variety of components, each of which must be taken into account throughout the process of developing a Digital Twin (object complexity). The basic objects comprise i) inputs such as seeds, feed, fertilizers, or pesticides; ii) throughputs such as objects in production (such as growing crops or animals); and resources such as fields, stables, machinery, and personnel; and iii) agricultural output such as harvested (lots of) crops, animals ready to be killed, *etc.*; and iv) throughputs such as objects in production (such as growing crops or animals); and v) (such as growing crops or animals). Digital Twins with a higher level of granularity, such as those that go down to individual plants or animals, would be more valuable; yet, it would be more expensive to develop such Digital Twins due to the intricacy of their design. In the scenario of fine granularity, one of the most significant issues is to manage the interdependences that exist between different granularity levels of (sub) Digital Twins. This is one of the most essential challenges.

Third, farms are part of a dynamic network and share data with a wide variety of stakeholders, including customers, input suppliers, farmer cooperatives, consultants, contractors, and certification and inspection organizations (network dynamics). It's possible that these stakeholders have access that's limited to a farmer's Digital Twins. This calls for compatible systems that allow trusted outsiders to view restricted portions of Digital Twins. On the other hand, external stakeholders have the ability to improve farm Digital Twins by contributing a vast array of third-party archives. Examples of these types of archives include historical and forecasted meteorological data, satellite data, results of soil-, water-, and air-analysis, and so on. The proper processes should be in place to dynamically incorporate these data into farm Digital Twins.

The use of digital twins represents a new stage in the evolution of "smart farming." It improves upon previously developed technologies, particularly those pertaining to precision farming, the Internet of Things, and simulation. As a direct result of this, there are several applications in the field of agriculture, but these applications are not always characterized as Digital Twins. The vast majority of these applications, however, are still rather simple kinds of Digital Twins. For instance, they may center on digital representation within a cloud dashboard. More complex applications, such as those with predictive and prescriptive capabilities over the entire product lifecycle, are still in the prototype phase.

## APPLICATION OF DIGITAL TWINS IN SMART FARMING

A computer depiction of a physical object that is accurate to the point that it may be considered its "twin" may be put to use to perform remote monitoring of the 'object.' For instance, a Digital Twin of a cow may notify the farmer of bad health without the farmer having to physically inspect the animal. This would save the farmer time and effort.

To be more explicit, the following are the requirements for a digital twin:

• Individual: It must be able to represent a particular instance, such as "Daisy, the cow" rather than "a cow in general."
• Near real-time: This also indicates that the Digital Twin should be "always on" or accessible for the same amount of time as its counterpart in the actual world will be around.
• Data-informed: It is necessary for it to be updated using a digital measurement of the object that exists in the physical world, such as a soil moisture metre or frequent satellite observations.
• Realistic: The Twin ought to be a substitute for the object that exists in the actual world that is as realistic as possible.
• Actionable: Information obtained from the real-world twin has to have the potential to lead to an action in order for it to be considered "actionable."

## USE CASES

• **Livestock Management:** Traditional applications for digital twins include herd management and supply chain optimization. Sensors on or in the cow (*e.g.*, accelerometers and boluses), at feeding and milking stations, and elsewhere (*e.g.*, weighbridge) collect a continuous stream of data. This data is then subjected to additional analysis in order to provide an indicator of the milk or meat production efficiency as well as the general health of the animal. After that, the farmer will be able to take the necessary steps [22]. Even if the activity that is carried out is not very unique, the information that is obtained from the

Digital Twin will enable an earlier and maybe more effective response, which will improve the outcomes for the animals.

- **Arable Farming**: In a farming setting, managers would often use digital duplicates of their fields to assess crop health. It would include a report on the existing qualities of the soil, water, crops, and any other relevant factors. In addition to that, they would provide support to the future predictions of such assets [23]. This information would be used to guide management measures, such as resolving gaps in production, the risk of lodging, projected over or under production, and crop rotation plans. Additionally, this information would be utilised to inform crop rotation plans.

- **Indoor Farming**: One of the most potential contexts for Digital Twins is controlled growth conditions, such as indoor farming, aquaculture, and under glass. These situations make it simpler to gather enough data for analysis and determine what is going on. A producer would routinely check in with the Digital Twin in order to monitor production, make adjustments (often from a distant location), and devise a plan for the project's ongoing administration. Early Digital Twins may find success in the controlled atmosphere and relatively high value of indoor farming.

- **Improving Sustainability**: It is feasible to use digital twins in order to offer guarantees on the Natural Capital of an agricultural landscape [24]. This can be done through a variety of different methods. Satellite data shows a lot of potential because it eliminates the need for strategically placed sensors to gather evidence. Carbon, biodiversity, pollination, and water catchment services are all areas where digital twins could be useful in agriculture. It would tell us whether they are changing and, more importantly, whether the changes that are being caused by us are being caused by them. Whether they are changing and whether the changes that we are causing are being caused by them. If the appropriate investments are made in sensors, 4 models, and interfaces, each of these use cases can now be brought to fruition. These, however, wouldn't always fit our concept of a digital twin because they would rely on general agricultural, cattle, and other models. To truly act as a Digital Twin, these models must take in external input and adapt based on their findings in order to produce an accurate depiction of the twinning object [25].

## CONCLUSION

It is expected that high-tech start-ups or sophisticated agribusinesses with considerable research and development expenditures will be the ones to generate digital twins [26]. Therefore, it is possible that startups like Connectra, who have the necessary knowledge, technology, and financing, could produce Digital Twins. Public sector organisations and major agribusinesses with innovation resources are also possible originators; John Deere and BASF's Xarvio platforms are developing in this arena. It's probable, however, that these businesses will need assistance, which is good news for nimble startups [27]. The more rapid development of digital twins in other industries, such as smart cities and smart factories, raises the possibility that a delivery capability could come into agriculture from elsewhere, provided that they can acquire the domain understanding. Other industries, such as smart cities and smart factories, are advancing at a faster rate than agriculture. This is because Smart Cities and Smart Factories are two businesses that have been aggressively adopting Digital Twins.

Taking into account their role in the life cycle, six distinct Digital Twins are identified:

1. **Imaginary Digital Twins**: Conceptual entities that describe and imitate reference objects that are not yet connected to objects that physically exist in the real world; conceptual entities;
2. **Monitoring Digital Twins**: Digital representations of the state and behavior, in (near) real-time, of real-world physical things, including the trajectory of those objects;
3. **Predictive Digital Twins**: Digital projections of the future state and behavior of physical items made using predictive analytics and based on (near) real-time data collected from the physical twins;
4. **Prescriptive Digital Twins**: Digital things with intelligence added for the purpose of suggesting remedial and preventative measures for physical objects;
5. **Autonomous Digital Twins**: Function without the need for human interaction either on-site or remotely, and be able to totally and completely regulate the behavior of real-world objects;
6. **Recollection Digital Twins**: Keep a record of the entire history of physical items even after they have ceased to exist in the real world.

## FUTURE SCOPE

The use of digital twins is unavoidable in high-value production systems in agriculture, such as those for indoor production, the production of fresh produce, and the raising of high-value animals. The foundations of these systems are already beginning to take shape, and it's probable that a genuine digital twin will become a reality within the next few years. They also seem likely to arise for the continuous monitoring and reporting system for the larger agri-landscape: ELMs and the increased focus on carbon and natural capital accounting will be significant drivers here. However, owing to technological difficulties, this will take a long time to complete. Finding solutions to lower the total cost of generating and deploying digital twins is going to be an important future topic for research and development. This capacity might be brought to many more applications if the study in question is fruitful.

# REFERENCES

[1]    S.J.C. Janssen, C.H. Porter, A.D. Moore, I.N. Athanasiadis, I. Foster, J.W. Jones, and J.M. Antle, "Towards a new generation of agricultural system data, models and knowledge products: Information and communication technology", *Agric. Syst.,* vol. 155, pp. 200-212, 2017.
[http://dx.doi.org/10.1016/j.agsy.2016.09.017] [PMID: 28701813]

[2]    A. Kamilaris, and F.X. Prenafeta-Boldú, "Deep learning in agriculture: A survey", *Comput. Electron. Agric.,* vol. 147, pp. 70-90, 2018.
[http://dx.doi.org/10.1016/j.compag.2018.02.016]

[3]    A. Tzounis, N. Katsoulas, T. Bartzanas, and C. Kittas, "Internet of Things in agriculture, recent advances and future challenges", *Biosyst. Eng.,* vol. 164, pp. 31-48, 2017.
[http://dx.doi.org/10.1016/j.biosystemseng.2017.09.007]

[4]    S. Wolfert, L. Ge, C. Verdouw, and M.J. Bogaardt, "Big data in smart farming–a review", *Agric. Syst.,* vol. 153, pp. 69-80, 2017.
[http://dx.doi.org/10.1016/j.agsy.2017.01.023]

[5]    A.T. Balafoutis, B. Beck, S. Fountas, Z. Tsiropoulos, J. Vangeyte, T. van der Wal, and I. Soto-Embodas, "Smart farming technologies–description, taxonomy and economic impact", *Precision Agriculture: Technology and Economic Perspectives,* pp. 21-77, 2017.

[6]    S. Boschert, and R. Rosen, *Digital twin—the simulation aspect*, 2016.
[http://dx.doi.org/10.1007/978-3-319-32156-1_5]

[7]    M. Grieves, and J. Vickers, *Digital twin: Mitigating unpredictable, undesirable emergent behav-ior in complex systems*, 2017.

[8]    S. Fountas, G. Carli, C.G. Sørensen, Z. Tsiropoulos, C. Cavalaris, A. Vatsanidou, B. Liakos, M. Canavari, J. Wiebensohn, and B. Tisserye, "Farm management information systems: Current situation and future perspectives", *Comput. Electron. Agric.,* vol. 115, pp. 40-50, 2015.
[http://dx.doi.org/10.1016/j.compag.2015.05.011]

[9]    M.E. Porter, and J.E. Heppelmann, "How smart, connected products are transforming com-petition", *Harv. Bus. Rev.,* vol. 92, no. 11, pp. 64-88, 2014.

[10]    M.W. Grieves, "Product lifecycle management: the new paradigm for enterprises", *Int. J. Prod. Dev.,* vol. 2, no. 1/2, pp. 71-84, 2005.
[http://dx.doi.org/10.1504/IJPD.2005.006669]

[11]    A. Redelinghuys, A. Basson, and K. Kruger, "A six-layer digital twin architecture for a manu-facturing cell", *Proceedings of SOHOMA 2018,* 2019 pp. 412-423.

[12]    K. Lepenioti, A. Bousdekis, D. Apostolou, and G. Mentzas, "Prescriptive analytics: Literature review and research challenges", *Int. J. Inf. Manage.,* vol. 50, pp. 57-70, 2020.
[http://dx.doi.org/10.1016/j.ijinfomgt.2019.04.003]

[13]    C.N. Verdouw, A.J.M. Beulens, H.A. Reijers, and J.G.A.J. van der Vorst, "A control model for object virtualization in supply chain management", *Comput. Ind.,* vol. 68, pp. 116-131, 2015.
[http://dx.doi.org/10.1016/j.compind.2014.12.011]

[14]    K. Blum, E.R. Braverman, J.M. Holder, J.F. Lubar, V.J. Monastra, D. Miller, J.O. Lubar, T.J. Chen, and D.E. Comings, *The reward deficiency syndrome: A biogenetic model for the diagnosis and treatment of impulsive, addictive and compulsive behaviors*, 2000.

[15]    C.G. Sørensen, S. Fountas, E. Nash, L. Pesonen, D. Bochtis, S.M. Pedersen, B. Basso, and S.B. Blackmore, "Conceptual model of a future farm management information system", *Comput. Electron. Agric.,* vol. 72, no. 1, pp. 37-47, 2010.
[http://dx.doi.org/10.1016/j.compag.2010.02.003]

[16]    L. Litavniece, S. Kodors, R. Adamoniene, and J. Kijasko, "Digital twin: an approach to enhancing tourism competitiveness", *Worldw. Hosp. Tour. Themes,* vol. 15, no. 5, pp. 538-548, 2023.
[http://dx.doi.org/10.1108/WHATT-06-2023-0074]

[17]    H. Fraga, A.C. Malheiro, J. Moutinho-Pereira, and J.A. Santos, "An overview of climate change impacts on European viticulture", *Food Energy Secur.,* vol. 1, no. 2, pp. 94-110, 2012.
[http://dx.doi.org/10.1002/fes3.14]

[18]    P. Batáry, L.V. Dicks, D. Kleijn, and W.J. Sutherland, "The role of agri□environment schemes in conservation and environmental management", *Conserv. Biol.,* vol. 29, no. 4, pp. 1006-1016, 2015.
[http://dx.doi.org/10.1111/cobi.12536] [PMID: 25997591]

[19]    G. Dyck, E. Hawley, K. Hildebrand, and J. Paliwal, "Digital Twins: A novel traceability concept for post-harvest handling", *Smart Agricultural Technology,* vol. 3, 2023.100079.
[http://dx.doi.org/10.1016/j.atech.2022.100079]

[20]    S. Neethirajan, and B. Kemp, "Digital twins in livestock farming", *Animals (Basel),* vol. 11, no. 4, p. 1008, 2021.
[http://dx.doi.org/10.3390/ani11041008] [PMID: 33916713]

[21]    E. Glaessgen, and D. Stargel, "The digital twin paradigm for future nasa and us air force vehicles," in 53rd AIAA/ASME/ASCE/AHS/ASC structures, structural dynamics and materials conference 20th AIAA/ASME/AHS adaptive structures conference 14th AIAA, p. 1818, 2012.

[22]    J. Srai, and E. Settanni, Supply chain digital twins: Opportunities and challenges beyond the hype," 2019.

[23]    Z. Kovacs, J-M. Le Goff, and R. McClatchey, "Support for product data from design to production", *Comput. Integrated Manuf. Syst.,* vol. 11, no. 4, pp. 285-290, 1998.
[http://dx.doi.org/10.1016/S0951-5240(98)00026-3]

[24]    M. Shafto, "Modeling, simulation, information technology and processing roadmap. washington, dc, usa: Nasa," 2012.

[25]    M. Philpotts, "An introduction to the concepts, benefits and terminology of product data management", *Ind. Manage. Data Syst.,* vol. 96, no. 4, pp. 11-17, 1996.
[http://dx.doi.org/10.1108/02635579610117467]

[26]    B. Schleich, N. Anwer, L. Mathieu, and S. Wartzack, "Shaping the digital twin for design and production engineering", *CIRP Ann.,* vol. 66, no. 1, pp. 141-144, 2017.
[http://dx.doi.org/10.1016/j.cirp.2017.04.040]

[27]    R. Sudarsan, S.J. Fenves, R.D. Sriram, and F. Wang, "A product information modeling framework for product lifecycle management", *Comput. Aided Des.,* vol. 37, no. 13, pp. 1399-1411, 2005.
[http://dx.doi.org/10.1016/j.cad.2005.02.010]

# CHAPTER 2

# Deep Learning Models for Prediction of Disease in Lycopersicum

**Nakatha Arun Kumar**[1,*] and **Sathish S. Kumar**[2]

[1] *BMS College of Engineering, Bangalore, India*

[2] *RNS Institute of Technology, Bangalore, India*

**Abstract:** In the Indian economy, agriculture is one of the main economic resources. Unemployment is one of the biggest problems. In many areas, agriculture offers about 60% of jobs. Various roles and responsibilities are involved in the agricultural sector including farmers, farm equipment operators, technical specialists, and more. There are a number of opportunities available in the agricultural sector. In developing countries, unemployment has decreased significantly since agriculture accounts for 20% of GDP, and agriculture can yield much greater benefits. Crops are classified into commercial and non-commercial crops. Most people can access and use non-commercial crops in their daily lives. Tomatoes are considered non-commercial crop. The edible berry of the species, Solanum lycopersicum, is commonly known as the tomato plant. Tomato plants are affected by many different diseases and because of these diseases, the losses are heavy. The disease can affect leaf, stem, or tomatoes. Initially, the disease is observed in the leaves, and eventually the disease worsens. Anticipating disease at an early stage is a major concern and therefore preventive measures are taken to obtain good crop quantity and quality. Applying deep learning techniques to disease detection has many advantages over other machine learning techniques. Deep learning is a part of machine learning techniques that helps train computers to do things that are natural to humans.

**Keywords:** Convolutional neural network(CNN), Deep learning(DL), Machine learning (ML), Neural network.

## INTRODUCTION

In India, the agriculture's contribution was 17% and has increased to 20%. Agriculture plays an important role in the Indian economy. Tomato is a non-commercial crop grown in India. Solanum lycopersicum is the species name of the tomato plant. Many diseases affect crops and early detection of these diseases is one of the challenges. Manual examination is tedious because it takes a lot of

* **Corresponding author Nakatha Arun Kumar:** BMS College of Engineering, Bangalore, India;
E-mail: arunnakhate@gmail.com

**S. Gowrishankar, Hamidah Ibrahim, A. Veena, K. P. Asha Rani & A. H. Srinivasa (Eds.)**

time. Integrating technology into early disease detection is provocative as there are many different types of diseases. Deep learning has many different functions that can be applied to detect disease and compare it, as it affects not only the economy but also the yield and quality of crops. Many factors affect tomato harvest and yield loss due to disease. Many factors damage agricultural crops, and diseases caused by bacteria and fungi, *etc.* are the common reasons. The application of deep learning techniques increases the accuracy of disease prediction and detection. In machine learning, deep learning is an advanced method that works like the human brain through the application of neural networks. Neural networks are part of deep learning. The advantage of deep learning is that it can automatically acquire features through image models. CNN is widely used in image analysis. CNN architecture can combine many different functions to achieve high accuracy. Traditional methods are applied to disease detection in machine learning. These traditional methods have limitations, therefore, to overcome the problem of the traditional method, a hybrid model has been proposed.

## RELATED WORK

The deep learning technique has attracted the attention of researchers because the deep learning CNN model automatically learns features. Research aspirants are deploying deep learning models to predict diseases in various plants—trigger functions like VGG, Unet, Alex Net, *etc.* The CNN model is applied to predict and detect diseases. As a result, accuracy has been increased in detecting diseases in tomato plants. Prediction and detection of plant diseases using their leaves is an active area of research and development. Plants are essential to life on Earth, providing medicine, nutrition, wood, and natural resources. Plant identification is a difficult task, even for experienced taxonomists and biologists. The color, texture, and shape of a leaf, all must differ in order to be distinguishable. However, due to seasonal color variations, shape and texture are the only relatively stable features that can be used. Automated plant disease detection algorithms and techniques are promising solutions to save many plant species from extinction. Significantly, manually maintaining a database of millions of crops is not an easy thing. This is where automatic plant disease detection comes in and helps researchers identify plants and store their information in a database.

Akash *et al.* [1], proposed a disease classification model for sunflowers, some of which are listed as late blight, cheese blight, alternate blight, and verticillium wilt. The merged model is applied in a pre-trained model stack *i.e.*, Mobile Net and VGG 16. The analysis and identification of plant diseases at an early stage is the most essential need to increase agricultural productivity, the agricultural sector plays an important role in the Indian economy. Early prediction of disease benefits agriculture by maintaining crop quality and quantity by saving crops and avoiding disease. Timely investigation of additional crop diseases is essential for better agriculture. An association deep learning model was introduced for disease classification, with four disease categories integrated into the data set. The stacker is used to combine VGG-16 and Mobile Net. The stacked set comparison with many other models was performed and the final model achieved 89.2% accuracy on the dataset. The model works well when compared to other models.

Muhammad Sufyan Arshad *et al.* [2], proposed transfer learning to identify diseases in maize, tomato, and potato. ResNet50, a pre-trained model, was integrated with the transfer learning method. The results were compared with VGG16 and MCNN. This model was built from scratch because 16 different plant diseases were identified by the model. This model has high performance with an accuracy of about 98°.

Bincy Chellapandi *et al.* [3], through a Plant Village dataset covering a wide range of interrelated diseases, reviewed about 38 plant diseases. Different pre-trained models were used, namely VGG16, VGG19, ResNet50, Inceptionv3, MobileNET, and DenseNet, and were combined for analysis. Dense Net achieved the best results and the accuracy was around 99%.

Xulang Guan *et al.* [4] used four pre-trained models of Inception, ResNet, Inception, and Dense-net in conjunction with the CNN model. The analysis was performed on different images of plants, as approximately; 36,258 images were used for the experiment. The dataset is divided into two parts, the training and validation set. A total of 31718 images are divided into the train and test model. An implementation that includes stacking a set has an accuracy rate of about 87%. The accuracy rate is increased by combining different pre-trained models by stacking.

Faizan *et al.* [5], performed a comparative analysis where traditional CNN methods were identified as techniques aimed at detecting plant diseases. The proposed model has an accuracy of 97.98%, which is the best test accuracy of the transformer model. Plant blight is a disease that has spurred the development of foliar disease prediction methods. This domain also allows early disease detection in plants. Convolutional Neural Networks (CNNs) are combined with various architectures, especially the "ResNet" architecture by using augmented data sets, covering a wide variety of diseases, *e.g.* healthy and unhealthy. According to performance, the Deep Learning technique has shown very good performance. Diseases are divided into two categories healthy and unhealthy. As a result, the system offers better performance than other models as the performance evaluation is compared.

Sammy V *et al*. [6], discussed the dataset that included 35,000 images and 9 groups related to different diseases related to Solanum Lycopersicum *i.e.*, tomato plants, 4 different diseases related to vines, 4 different diseases related to maize, 4 different diseases related to maize and apple, and 4 different diseases associated with sugarcane. The CNN model is implemented, and the data set is divided into rain and test. This application has been integrated with data enhancement techniques to improve efficiency. The enhancement is applied to the image dataset by rotating the images to different degrees, flipping them, rotating them, and shifting them in various shapes horizontally and vertically. Among the various optimization techniques, Adam Optimizer is incorporated. The model is trained in 32 batches, and the number of epochs is 75 epochs.

Md. Arifur Rahman *et al*. [7] described segmentation as one of the main goals in establishing threshold and morphological activities. The deep neural network was applied as the classifier and the method achieved an accuracy of 99.25. Among the different segmentation techniques, RGB thresholding and morphological activity were applied. The aim is to improve the accuracy and size of feature vectors as well as take into account the classification. This work can be extended by combining a larger data set that includes many different diseases of different plant species.

Draško Radovanović *et al*. [8], presented classical algorithms that support vector machines, K-nearest neighbors, and faster convolutional and convolutional neural networks. The implementation is done through Python, scikit-learn, and Keras libraries. Robert G. de Luna *et al* [9] described as tomato plants were used for disease detection and disease severity in Diamante Max, Target Spot, Phoma Rot, and Leaf Miner plants. The dataset consists of approximately 4923 healthy tomato leaf images, which were obtained from a variety of sources. In specific environments where images are captured and repositioned, including healthy and unhealthy leaves, the model is trained to predict detected defects. The model achieved an accuracy of 95.75%. The DCNN architecture [10, 11], AlexNet, and GoogleNet consist of five convolutional layers and three connected layers. The results show that DenseNet performs best on the test data with 99% accuracy.

Deep learning belongs to machine learning and is also recognized as a subcategory of machine learning. Deep learning is widely used for image analysis, mainly in disease analysis. Hierarchical features are studied automatically without combining feature design for extraction and classification.

The CNN process flow, as shown in Fig. (**1**), is aimed at disease classification and works better than machine learning techniques. Convolutional neural network was chosen as the basic model because it works well and CNN is the best technique among others. CNN is applied not only to large data sets but also to small data sets as the best and most efficient image classification tool. CNN is an underlying architecture that is considered the underlying model. The CNN model is considered the basic model, which is applied to many forms of pre-training. Convolutional Neural Networks (CNN) work well and successfully with good classification results in deep learning platforms. In this study, visualization was reviewed and analysis was performed based on various types of neurons and layers. The CNN model can identify colors and textures to predict specifically different diseases during the diagnostic process, much like human decision-making.

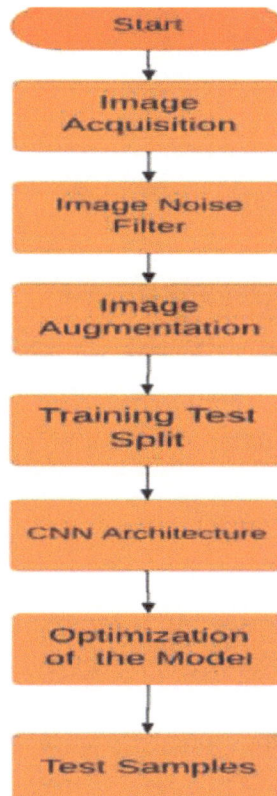

**Fig. (1).** CNN process flow.

## Pretrained Models

Various applications of artificial intelligence are attracting a lot of attention, of which deep learning has become an important tool. In the field of computer vision, image analysis, text analysis, speech recognition, *etc.*, the use of deep learning models and the results obtained from these models have had a significant impact on the research field. The goal of image classification is to classify a particular image according to a set of possible categories. In analytics, it is easy to draw conclusion about predictions. Many architectures have been introduced in recent years. Several deep learning architectures have been defined and redefined to address critical problem-solving. The refinement of architecture is performed according to the change of the hyperparameters of the classes. The deep learning architectures are listed below:

AlexNet

VGG16

GoogleNet

ResNet

Inception

AlexNet

The best-known architecture is the Alex network architecture. This architecture was introduced by the Imagenet Large-scale Image Recognition Challenge. The Alex Net architecture consists of different layers. Architecturally, it has eight layers, out of these, three are meant to be fully connected and five are convolutional layers. The maximum pooled layer is connected to the first two convolutional layers whereas, the other layers are connected to the fully connected layers.

**VGGNet**

The VGG architecture is cumbersome compared to the AlexNet architecture, consisting of many large (3×3) filters. VGGNet is applied to a convolutional neural network and applied to the Image Net 2014 challenge. This network has an accuracy of 92.7 applied to the ImageNet dataset.

**GoogleNet**

In the prediction domain, this network is responsible for defining a new modern state. This first version was brought into the field in 2014 when the Google team introduced the Google network.

**ResNet**

Residual mapping technique was introduced, and this network used deep learning for image recognition in 2015 and was named ResNet-Residual Network.

## Inception

The Inception model was introduced to overcome the computational aspect, as it was developed to solve the problem of computational efficiency. Different versions of the Inception network specific to CNN were introduced, and different versions of Inception *i.e.*, Inception 1, Inception2, and Inception3 were introduced.

## Methodology

1. Training Data

10599 images belonging to 9 classes.

2. Testing Data

2656 images belonging to 9 classes.

3. Loss = categorical cross entropy

4. Optimizer = RMSprop - Root Mean Square Propagation.

5. Currently trained for 10 Epochs, the below metrics are on the testing dataset.

The learnable parameters were used in the convolutional neural operation using the kernel, with the learnable parameters W and b passed to x pixels for each image to obtain the result. Core movement is of two types, pixel-based skip pixels, and stride-defined pixels. Automatic Feature Extractor is a convolution layer that extracts patterns for image classification. The original edges of the image will be examined later and these features are combined to form complex features. A non-linear activation layer was integrated and a rectified linear unit was integrated. The obtained results were fed to the SoftMax class to generate a probability score for each class.

## RESULTS

The Solanum lycopersicum dataset included 10 different diseases in tomato plants, including images of healthy tomato leaves. For disease prediction, the dataset included images of healthy leaves. Pictures of tomato leaves were collected from many different regions. The data set was divided into a training set and a test set, for the training model, 80% was used and 20% for the test model. The CNN model achieved the highest accuracy.

## CONCLUDING REMARKS

The performance of the classification models has improved. The classifiers performed efficiently on the validation dataset. Accuracy was 88%, 77%, 81%, 72%, and 28% for CNN, InceptionNet, ResNet, VGGNet, and AlexNet, respectively. The concluded part is about the accuracy of the proposed models, The model was improved by the Meta model and compared with the hybrid model implemented in combination with an optimization technique.

## REFERENCES

[1]    A. Sirohi, "A hybrid model for the classification of sunflower diseases using deep learning", *Second International Conference on Intelligent Engineering and Management (ICIEM),*. London, United Kingdom, 2021.
[http://dx.doi.org/10.1109/ICIEM51511.2021.9445342]

[2]    A. Muhammad Sufyan, "Plant disease identification using transfer learning", *International Conference on Digital Futures and Transformative Technologies (ICoDT2),*. Islamabad, Pakistan, 2021.
[http://dx.doi.org/10.1109/ICoDT252288.2021.9441512]

[3]    B. Chellapandi, "Comparison of pre-trained models using transfer learning for detecting plant disease", *International Conference on Computing, Communication, and Intelligent Systems (ICCCIS),* Greater Noida, India, 2021.
[http://dx.doi.org/10.1109/ICCCIS51004.2021.9397098]

[4]    X. Guan, "A novel method of plant leaf disease detection based on deep learning and convolutional neural network", *6th International Conference on Intelligent Computing and Signal Processing (ICSP),* Xi'an, China, 2021.
[http://dx.doi.org/10.1109/ICSP51882.2021.9408806]

[5]    Faizan, "Plant disease detection using deep learning algorithms: A systematic review", *Scie. Tech.,* vol. 4, no. 2, 2023.

[6]    V. Sammy, "Plant leaf detection and disease recognition using deep learning", *IEEE Eurasia Conference on IOT, Communication and Engineering (ECICE),* Yunlin, Taiwan, 2019.

[7]    Md. Arifur Rahman, "Improved segmentation approach for plant disease detection", *1st International Conference on Advances in Science, Engineering and Robotics Technology,* Dhaka, Bangladesh, 2019.

[8]    D. Radovanović, "Image-based plant disease detection: A comparison of deep learning and classical machine learning algorithms", *24th International Conference on Information Technology (IT),* Zabljak, Montenegro, 2020.

[9]    R.G. de Luna, "automated image capturing system for deep learning-based tomato plant leaf disease detection and recognition", *TENCON 2018 : 2018 IEEE Region 10 Conference,* Jeju, Korea (South), 2018.
[http://dx.doi.org/10.1109/TENCON.2018.8650088]

[10]   Y. Haseeb, "Robust optimization of mobile net for plant disease classification with fine tuned parameters", *International Conference on Computer, Control, Informatics and its Applications,* Islamabad, Pakistan 2019.

[11]   R. Sandra Yuwana, "Multi-Condition Training on Deep Convolutional Neural Networks for Robust Plant Diseases Detection", International Conference on Computer, Control, Informatics and its Applications, 978-1-7281-5540- 1.

# CHAPTER 3

# A Smart Hydroponics System for Sustainable Agriculture

**Supriya Jaiswal[1,*], Gopal Rawat[1], Chetan Khadse[1]** and **Sohit Sharma[1]**

*[1] National Institute of Technology, Hamirpur (H.P.), Himachal Pradesh 177005, India*

**Abstract:** The agriculture sector not only contributes to the nation's economy but also serves as an important element in foreign exchange and trade markets. With the advancement in technology, robots, drones, satellite imagining, IoT, wireless sensor networks, machine learning, big data analytics, and unmanned aerial vehicles (UAV) are being deployed to manage, monitor and control agricultural chores. However, the farmers are unable to meet the increasing urban food demand with limited cultivable land availability. Thus, to solve this issue, hydroponic farming is opted for in several parts of the world. It is a soil-free and nutrient-rich water medium for agriculture, which is increasingly opted for by the urban population. Hydroponic farming has been vastly explored in the context of urban farming, where land, water, time, and labour are required in a limited amount, yet productivity is far better compared to traditional agricultural methods.

It has been recently adopted in urban sections in India due to restricted movement in COVID-19 pandemic situations to fulfil basic food requirements. However, hydroponic farming has shortcomings such as higher initial cost, the possibility of complex nutrient discharge problems, the energy requirement for the creation of microclimatic conditions, fertigation and effluent treatment and pretrained skilled labour. In order to resolve these issues, a smart hydroponic farming architecture is discussed, which reduces human intervention and water wastage using wireless sensor networks and IoT. In order to successfully and efficiently implement the agricultural supply chain, machine learning algorithms and data mining techniques are utilized from the production to inventory storage stage. The following sections deal with a brief introduction to hydroponic farming, its architecture and components, and future opportunities regarding the field of automated hydroponic farming.

**Keywords:** Hydroponic farming, Food security, Nutrient media, Internet of things, Wireless sensor network, Machine learning algorithms.

---

* **Corresponding author Supriya Jaiswal:** National Institute of Technology, Hamirpur (H.P.), Himachal Pradesh 177005, India; E-mail: supriya@nith.ac.in

S. Gowrishankar, Hamidah Ibrahim, A. Veena, K. P. Asha Rani & A. H. Srinivasa (Eds.)

## INTRODUCTION

In order to meet the increasing food demand, the agro industries and food markets have initiated different farming practices across the globe. In the current scenario of Indian agriculture, various types of farming are deployed, *e.g.*, organic farming, precision farming, hydroponics, greenhouse farming, micro -irrigation-based farming, *etc.*

## TYPES OF FARMING

i. **Organic Farming:** It is an integrated farming approach which aims for the sustainability and improvement of soil fertility by harnessing fertilizers of organic composition such as compost. It restricts the usage of synthetic pesticides, synthetic fertilizers, and growth hormones. There are different benefits of these integrated farming practices, such as ecological sustainability, recycling, energy saving, increase in productivity and profitability.

ii. **Precision Farming:** It is based on observing, measuring, and responding to inter- and intra-field crops and their environmental variability. The objectives of precision farming are: (i) to closely match the fertilizer inputs, (ii) to limit leaching for environmental protection, and (iii) improved irrigation management.

iii. **Hydroponic Farming:** In this method of farming, the plant root is submerged in the nutrient-rich water solution. The composition of nutrients is controlled by continuously monitoring the crop growth cycle. Hence, the crop grows in the optimized microclimatic conditions in the limited space to produce healthy crop production. This type of farming establishment is costly, although it a is viable solution for the urban population to meet their food demands and reduce the crop burden on farmers. This technique also helps to reduce fertilizer, pesticide usage and abolish soil-borne diseases.

iv. **Greenhouse Farming:** Greenhouse farming is a well established and popular means of farming in colder regions. It is made of framed structures covered with transparent/translucent material in which the crop grows in controlled environmental conditions. The parameters which can be controlled using greenhouse farming are temperature, light, carbon dioxide levels, humidity, water, nutrient, and pests. Thus, it allows the off-season production of high-value crops and flowers in cold climatic areas where outdoor production is not possible. However, this type of farming requires high installation and operating costs. Moreover, pollination gets poor in the greenhouse structure.

v. **Micro-irrigation-based Farming:** Apart from nutrients, fertilizers, and pesticides, water plays an important role in plant healthy growth and production. Micro-irrigation provides water to the plant root zone in low volume and frequent intervals under low pressure to maintain its healthy growth. In order to maintain automated and controlled irrigation according to crop needs, sophisticated irrigation methods have been introduced in farming in recent years, *e.g.*, micro-sprayers can be used for large surface areas, micro bubblers can be used on larger plants which reduces the rate of evaporation compared to sprinklers. Drippers are applied to supply water directly to the soil and can be used for plants kept in containers. It minimizes evapotranspiration.

In addition to these farming practices, the other types of farming methods are conservation agriculture which emphasizes crop rotation, aquaponics where the nutrient mix water can be reused for fish cultivation and aeroponics where showering/splashing the nutrient mix mist using nozzles on the plant (without soil medium) helps in plant development.

## Hydroponic Farming

The term Hydroponic is extracted from the Greek words, "hydro" means water and "ponos" referring labour. This implies the farming method where water is used as a working agent for ensuring plant growth. The diverse variety of plants requires different soil media for healthy growth. Some plants require heavy water to be retained by soils like clay soil, others may require sandy soil to drain out standing water easily, and other varieties of crops need loamy soils which neither hold nor drain the water completely. So, these demands automatically seize the cultivation of diversity. In this scenario, hydroponics emerges as a viable method for the cultivation of a wide variety of crops without soil in only a nutrient-rich water medium. Thus, the roots of crops do not have to search for nutrients in the soil and readily access all essential nutrients promoting the rapid growth of crops. In addition to this, soil farming also provides several benefits such as efficient usage of water and nutrients, no chemical fertilizers, herbicides and pesticide usage, high production in less space, time and less environmental impacts.

There are different types of hydroponic systems which are broadly categorized into two groups (i) solution culture and (ii) medium culture, as shown in Fig. (**1**). In solution culture, the roots of the plants are kept in an air or water medium mixed with nutrients, whereas in medium culture, the root of the crop grows in solid medium, such as rockwool and cocopeat.

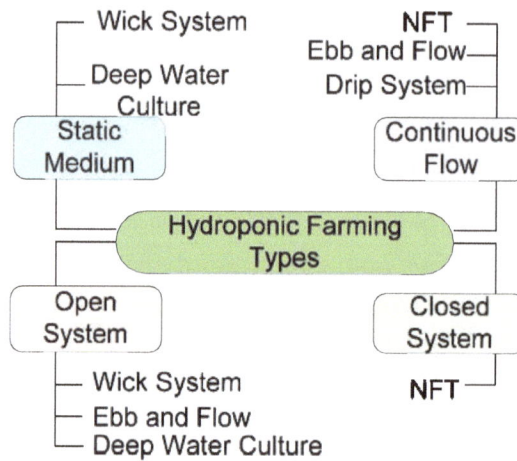

**Fig. (1).** Categorization of Hydroponic Farming.

i. Solution culture relies upon three different ways of growing the crop in solution:

   a. **Static solution culture:** As the name suggests, here, the crops are grown in containers of nutrient solution. The aeration in such an arrangement can be provided similar to aquaria. The container is covered with foil or plastic to exclude light, which eliminates the formation of algae. The electrical conductivity meter is used to measure the nutrient levels in the nutrient solution, and fresh nutrient soil is added when the concentration drops below a certain level.

   b. **Continuous flow culture:** In this method, the nutrient solution continuously flows through the roots. The nutrient film technique is one of the examples of continuous flow culture, where a narrow stream of nutrients mixed with water is circulated through the crop roots constantly. The growing tray is kept tilted at some slope so that the excess nutrients get back into the nutrient reservoir. Aeration is provided using air pumps and air stones similar to an aquarium to ensure healthy plant growth.

   c. **Aeroponics:** In this system, the roots of crops are suspended in container and nutrient rich mists are sprayed or splashed on the crop roots' surface using nozzles. The major demerit of this method is that it is hard to ensure that complete root sections are sprayed with nutrient mist.

i. **Medium culture:** In this method, there are two variations for each media (a) passive sub-irrigation and (b) top irrigation.

   a. **Passive irrigation:** In this method, the crops are grown in a porous medium. The plant roots transport water and nutrients by capillary action. This method reduces the possibility of root rot and additional ambient humidity.

b. **Top irrigation:** The nutrients are applied to the surface of the media. This procedure can be automated with pump time and drip irrigation or manually once per day.

There are several other popular hydroponics systems which can be opted for based on the types of crops to be grown. Fig. (**2**) shows the pictorial view of different types of hydroponic farming systems.

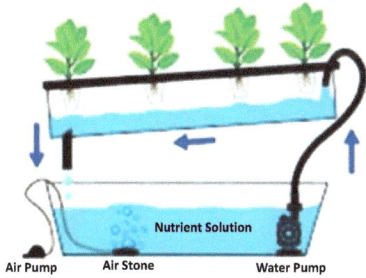

**(a) NFT Technique**

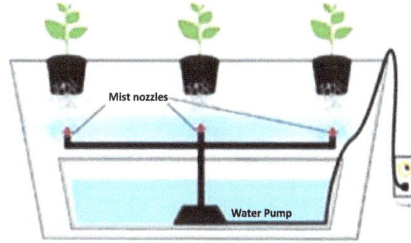

**(b) Aeroponics**

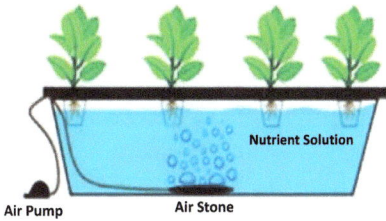

**(c) Deep Water Culture Technique**

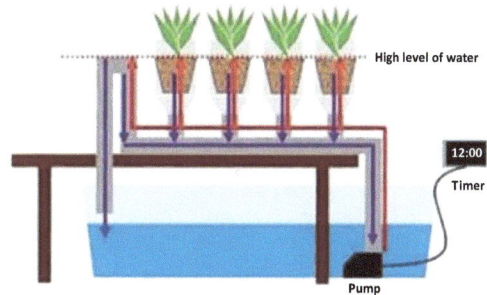

**(d) EBB and flow hydroponic system**

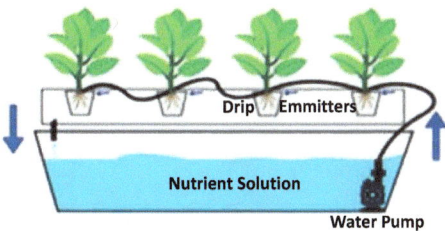

**(e) Drip System**

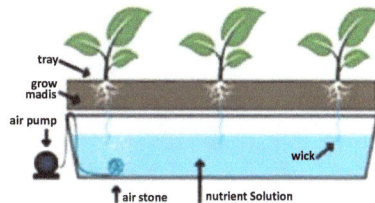

**(f) Wick System**

**Fig. (2).** Different types of Hydroponic Farming Systems.

a. **Deep water culture:** In this active water culture method, a floating platform holds the plant in nutrient solution. Air pumps are used to supply oxygen to roots. The plant growth is quite good because of the abundant availability of water and oxygen.

b. **EBB and flow hydroponic system:** In this method, a submerged water pump is used to flood the nutrient in the growing tray, and then the nutrient is drained back to the reservoir. When the timer shuts the pump off, the nutrient solution returns back to the reservoir. The pump operation depends on the plant growth cycle, humidity, and temperature.

c. **Drip system:** In this system, the nutrient solution is dripped in the root area of each plant using a drip line. This system is automated using a pump system and aeration. The excess solution is recycled into the nutrient reservoir in a recovery drip system, whereas in a non-recovery system, the excess nutrients are drained out. Thus, it recovers a precise watering cycle to give sufficient nutrients to the crop and the run off is kept minimum.

d. **Wick system:** It is a simple passive hydroponic system where a wick is utilized to draw the nutrient from the nutrient tanks to the crop growing medium. The growing media can be cocopeat, vermiculite, and perlite.

## Scope and Challenges

Growing crops in controlled environmental conditions is a quite complex method of cultivation. Before opting for hydroponics as a mode of farming, the grower should analyse critical aspects such as site selection, structure, hydroponic growing system, pest control, and market.

**Site selection:** One of the important factors for locating hydroponic farming is the light intensity and its duration over a particular site. Generally, at a reduction of 1% of light, the yield reduces by 1%. Another important factor is temperature. It is difficult to grow off-season crops in winter climatic zones or in extreme high temperature zones as it would require appropriate heating and cooling needs, respectively.

Moreover, the selected site should be pests and insects free. Growing crops in mild winter regions where insects flourish easily should not be chosen for hydroponic farming.

i. **Energy and water requirements:** Since most of the hydroponic system has a need for automated nutrient flow from the nutrient tanks to the growing plates and aeration through automated air pumps, thus energy is required to facilitate such systems. Water is a major factor in establishing hydroponic farming. The plant growth depends on the quality of water (*e.g.*, pH and electrical conductivity). These chemical properties determine the nutrient absorbability in the water that can be absorbed by plant roots through capillary action. Thus, water quality should be ascertained before establishing hydroponic systems.

ii. **Structure for environmental control:** Greenhouse is generally opted for maintaining heat in the growing area. Controlling the environmental condition requires energy and should be conserved once sunlight incidents in the greenhouse. In order to fulfil this purpose, thermal curtains can be used to reduce heat loss by 57%. A remote control centre with a computer can monitor all sensor data such as temperature, humidity, nutrient level, light, and $Co_2$ to keep the crop growth in check.

iii. **Selection of media culture system:** There are various types of media available for growing systems, such as diahydro, rockwool, cocopeat, perlite, sand, quartz, gravel, and brick. However, in India, cocopeat and perlite are the most commonly used media for hydroponic farming.

iv. **Pest control:** In order to reduce the pest, the drain solution is sterilized using heat treatment, ozone, and ultraviolet radiation. To control pests, integrated pest management is a viable solution. It detects insect problem, identifies crop disease, and prevent pests.

v. **Market:** Depending on the food demand in the market, the respective crops can be grown. Such specific crop production on a domestic or commercial scale can be accomplished using suitable hydroponic farming architecture.

## ADVANTAGES AND LIMITATIONS OF HYDROPONICS

Hydroponic farming has several advantages over conventional agriculture practices, as highlighted in Fig. (**3**). A few of the advantages are discussed below:

1. **Higher productivity-** As compared to conventional agriculture, hydroponics yield higher crop cultivation. The growth of the crop is far better as there is no mechanical obstruction to the roots to absorb nutrients.

2. **Soil-less medium-** Since hydroponic farming enables the crop to grow in a nutrient mix solution medium, the nutrients are readily available to the crop, hence their growth is quite appreciable compared to soil medium. Moreover, the diversity of different crops needed for various types of soil is completely eliminated in hydroponics.

3. **Less space requirement-** Hydroponic farming can be accomplished in small spaces like a roof to yield the same production as those grown in fields.

4. **Maintenance and operation-** Once the hydroponic farm is completely installed and automated, there is little need for human intervention to operate it, and the least maintenance is required.

5. **Water conservation:** Hydroponics uses $1/20^{th}$ amount of regular farm irrigation needs and the water lodging phenomenon never happened in this method. Thus, water is utilized more efficiently.

6. **Pests and disease problem-** As hydroponic farming does not rely on soil for crop growth, hence weeds and pest problems are non-existent.

7. **Control over the plant growth-**The plant rooting environment can be easily controlled and monitored using sensors and communication networks to detect and maintain temperature, humidity, light, pH, and electrical conductivity.

8. **Higher returns-** Plants that are off seasonal can be easily grown in hydroponic farms to fetch higher gains.

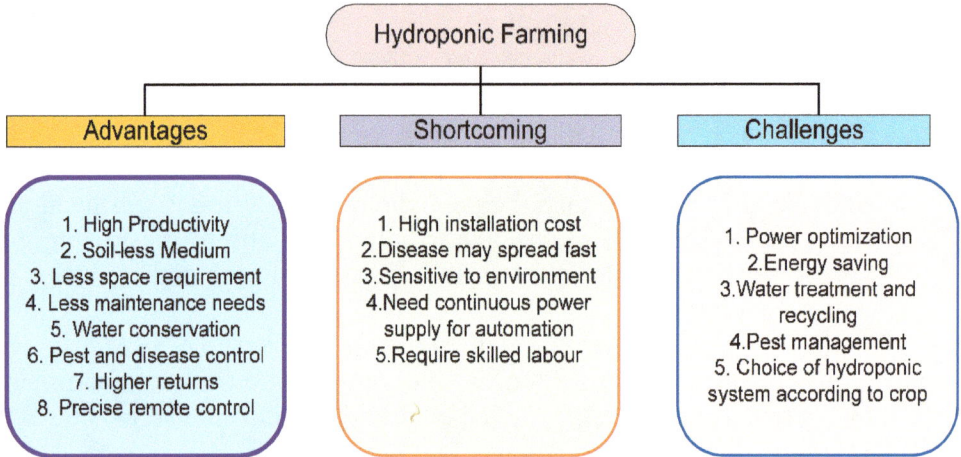

**Fig. (3).** Advantages, shortcomings and challenges faced in Hydroponic Farming.

The challenges faced in hydroponic system establishment are:

1. High installation cost.
2. Technical skill and knowledge are required to operate the hydroponic farm.
3. Since the same nutrient medium is used for growing the tray, any disease can easily spread to other plants and affect them.
4. As the crops grow under controlled microclimatic conditions, any adverse change in the environment can quickly worsen the condition of crops.
5. Hydroponics farming is mostly automated. Thus continuous supply should be maintained with the sensor network, and failure of supply can lead to limited oxygenation and nutrient deprivation, further leading to production loss.

## GOVERNMENT INITIATIVES AND RECENT RESEARCH TECHNOLOGIES TO SUPPORT FARMING

**Government initiatives**: Farming is the basic source of income and livelihood for 58% of the Indian population. Farming, forestry and fishing add Rs. 19.48 lakhs crore gross value addition in FY20, in which agriculture carries 17.85% shares. The food processing sector in India accounts for 32% of the country's food market. It is the 5th largest industry ranked in terms of production, consumption, and expected growth. From April 2000 to June 2021, according to the Department of Industry Promotion and Internal Trade, the Indian food industry has attracted an equity inflow of 10.43 Billion USD from foreign direct investment. With significant interest, India ranks 3rd in terms of agritech funding and a number of agritech start-ups. By 2025, India's agritech sector will witness a growth of 30-35 Billion USD.

The government has also planned several policies and schemes to help the farmer and agritech industries. Ministry of Aviation launched Krishi UDAN 2.0 scheme in October 2021 to give incentives for the movement of agricultural products by air transport. It will benefit the Northeast and tribal regions. Moreover, Pradhan Mantri Kishan Samman Nidhi Yojana (PM-Kisan) has been implemented, under which 284.48 Million USD is transferred to direct 10 million farmers' accounts to help them establish their farming needs. As per Union Budget 2021-2022, Rs. 4000 Crore was allotted towards implementing PM Krishi SinchaiYojana aimed at the development of irrigation sources.

The Ministry of Food Processing has been allocated Rs. 1,308.66 Crore in Union Budget 2021-2022. Government plans to triple the capacity of the food processing sector in India from the current 10% of agricultural produce and committed USD 793 million to Food Park under the Scheme of Agro-Marine Processing and Development of Agro-Processing (SAMPADA) cluster.

**Recent research technologies for hydroponics**: An intelligent hydroponic system where all process for growing crops is based on Bayesian Network was proposed having 84.53% accuracy and improved yield. This technique was capable of monitoring and controlling agricultural parameters like humidity, temperature, light intensity, water electrical conductivity (EC), and pH [1]. Machine learning based method for analyzing the growth dynamics of a plant and reducing energy input was developed to study the challenges encountered in smart farming [2]. Moreover, ML and image processing techniques were used to determine the freshness of plants grown in a hydroponic environment so as to distinguish between withered and fresh crops. With the use of a decision tree, the performance of the algorithm exhibited 98.12% accuracy [3]. The authors proposed k-Nearest Neighbour (KNN) and IoT to control nutrient flow, temperature, pH value, and dissolved solids using sensor data. This method achieved an accuracy of 93.3% [4]. A model based on IoT and linear regression with ML techniques was used to determine the quality of lettuce grown through hydroponic farming. The features used for analysis and control are humidity, light intensity, and temperature. Further, leaf area, total weight, and nitrate content were target variables [5].

Apart from the neural network and other intelligent techniques, some researchers also automated hydroponic farming using microcontrollers and communication networks. An intelligent hydroponic system, where real-time information can be transferred in AVR microcontroller and Lab VIEW for automated hydroponic cultivation, is proposed [6]. In a study [7], an approach based on IoT and mobile applications is proposed to automate hydroponic farming. The system monitors the temporal variations in various farming parameters. Moreover, farmers can be intimidated by unfavourable conditions through real-time measuring services.

In another research [8], the authors proposed IoT-dependent hydroponics for lettuce cultivation. Here, the system used for the LED lighting method is facilitated by IoT. Arduino microcontroller based intelligent data acquisition method was developed for the hydroponic system. The farming parameters were measured using a sensor which presented less percentage error [9]. A webpage was created for the hydroponic systems. All sensors and actuators were supplied through solar panels, and thus NFT hydroponic system utilized less power [10]. SWOT Analysis of Hydroponic farming is discussed in Fig (**4**).

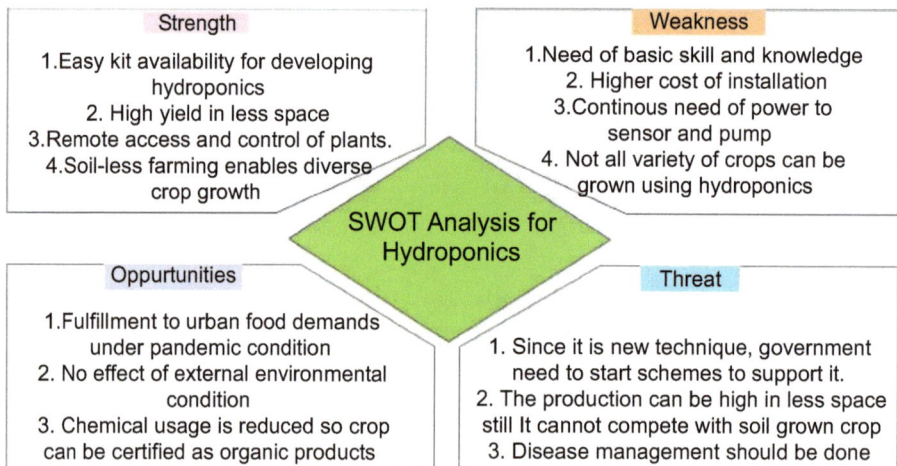

**Strength**
1. Easy kit availability for developing hydroponics
2. High yield in less space
3. Remote access and control of plants.
4. Soil-less farming enables diverse crop growth

**Weakness**
1. Need of basic skill and knowledge
2. Higher cost of installation
3. Continous need of power to sensor and pump
4. Not all variety of crops can be grown using hydroponics

SWOT Analysis for Hydroponics

**Oppurtunities**
1. Fulfillment to urban food demands under pandemic condition
2. No effect of external environmental condition
3. Chemical usage is reduced so crop can be certified as organic products

**Threat**
1. Since it is new technique, government need to start schemes to support it.
2. The production can be high in less space still It cannot compete with soil grown crop
3. Disease management should be done

**Fig. (4).** SWOT analysis for hydroponics.

# COMPONENTS AND STRUCTURE OF SMART HYDROPONIC FARMING

The structure and components of a hydroponic verticulture system are majorly constituted of different parts such as a supply system (conventional grid/ solar based), high grade PVC pipe-based NFT system consisting of seedlings, water and nutrient tank, automatic nutrient flow system consisting of the electrical solenoid valve, sensor network, microcontroller unit, IoT network for sensor data collection, yield monitoring and automated fertigation. Fig. (**5**) shows the schematic diagram of the proposed smart hydroponic farming system.

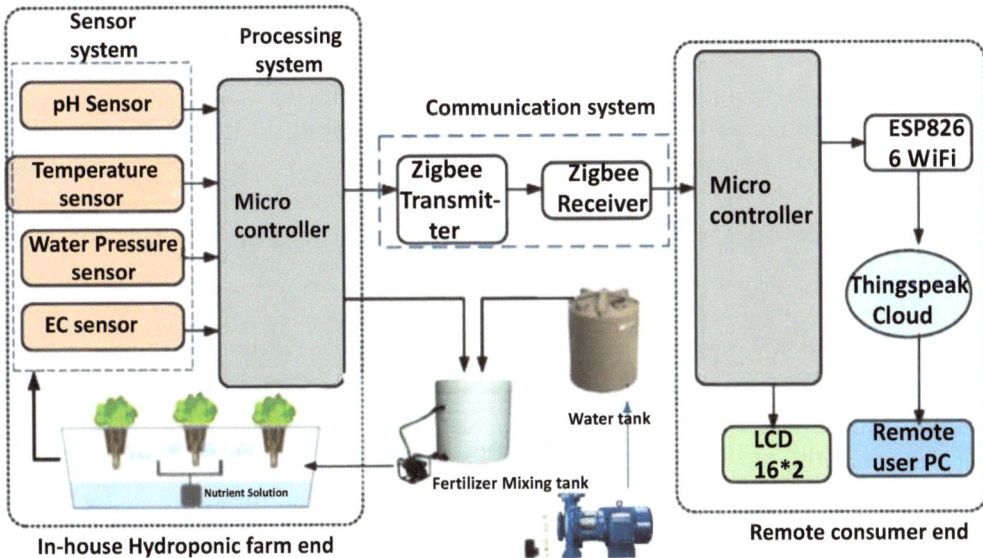

**Fig. (5).** Schematic diagram of the proposed smart hydroponics system.

## Supply System for Automated Water Pumping

The hydroponic fading is primarily dependent on proper nutrient mix flow, which can be accompanied regularly based on the data communication through the sensor network. All these farming activities can only be accomplished if the power supply to the water pump, solenoid, sensors, control unit, and IoT is maintained continuously. The power supply needs could be catered either by traditional means, *i.e.*, grid supply or by solar panels. Apart from this, battery banks are also provided to improve supply reliability. Installation of Solar PV panels makes the system more eco-friendly and cost-effective in the long run. To fetch the maximum power, MPPT is utilized in the solar pumping system. Both MPPT and motor control are carried out by a three-phase voltage source inverter. On the other hand, the bidirectional converters provide the facility to store energy in the battery banks. The IOT and WSN are powered by solar photovoltaic panels. The solar charge controller is provided to prevent the battery from overcharging and overvoltage.

## Sensor Network for Smart Farming

Smart sensor availability provides high yields in the minimum resource presence such as water, fertilizers, and seeds. It is easy to install and are user friendly in facilitating different functionalities. Sensors play a vital role in hydroponics agriculture right from data gathering to decision making more efficiently.

Smart sensors add innovation in hydroponic farms through automation and IoT based monitoring and control operation using digital communication networks. The data-driven technologies harness smart sensors to solve many issues of agricultural practices. Artificial intelligence, machine learning, and fuzzy inference based decision making, in addition to IoT enabled sensor-based machines, provide the farmer insight into different stages right from sowing to harvesting. The different types of sensors used in hydroponic farming are as follows:

1. **EC sensor:** In the hydroponic system, plant growth is governed by an appropriate amount of nutrient mixes like nitrogen, potassium, phosphorus, magnesium, calcium, zinc, copper, boron, sulphur and molybdenum. The strength of the nutrient solution is monitored using electrical conductivity (EC), which infers the conductivity value of nutrients when they dissociate into the water. The ideal conductivity level in the nutrient solution helps in the growth of healthy crops.
2. **pH sensor:** pH sensor is a photometric sensor that uses pH-sensitive reagents to analyze the solution pH levels. The plant roots' ability to absorb the nutrient mix depends on the pH level of the solution. For example, the appropriate acidic concentration of nutrient solution helps in the absorption of manganese, aluminium, and hydrogen, although it's excess can be detrimental to plant health. The low pH level decreases calcium and manganese absorption leading to nutrient deficiency. In a similar manner, alkaline pH increases the absorption ability of macronutrients and decreases phosphorus, zinc, copper, cobalt, and iron absorption. Thus, to make the nutrient available at the correct pH for plants, a pH sensor is compulsorily required.
3. **Ion Selective Electrode and Ion Meter:** Ion selective electrode works on the Nerst equation. It can detect individual cation and anion concentrations such as nitrate, ammonium, and potassium levels accurately. This way, the ion meter can control the individual ion concentration in the hydroponic nutrient solution.
4. **PAR Sensors:** These are photosynthetically active radiation sensors which measure the light intensity that is critical for plant growth. There are several photodiodes, such as Silicon (Si) and gallium arsenide phosphide (GAsP), that are used to measure light intensity in wavelength. Light sensor ensures the

proper light admittance required for crop growth, similar to greenhouse farming design.

5. **Co₂ measurement sensor:** These sensors are vital for crop growth and its monitoring is accomplished using $CO_2$ sensors such as NDIR $CO_2$ sensors which can be used for small scale hydroponic farms.

6. **Optical sensors:** Optical sensor senses the plant nutrients from the leaf reflectance spectra non-invasively. The shorter wavelength depicts the plant stressed conditions. These sensors ensure the proper light admittance required for crop healthy growth, similar to greenhouse farming design.

7. **Ultrasonic sensors:** The ultrasonic sensor checks the level of water in the tank to ensure regular flow of nutrient water mix, which is regulated by an electrical actuator (solenoid) valve and meter to monitor the volume of water. These actuators can be monitored and controlled using sensor networks and web applications remotely using IoT. Table **1** shows the different smart sensors used in hydroponic farming.

**Table 1. Smart sensors used in hydroponic agriculture environment.**

| Sensor | Products | Function | Sensor example | Manufacturer |
|---|---|---|---|---|
| pH sensors/ pH meter probes | S-pH-01-A CH-9328 TT-Hydro- pH ENV-4- -pH | Determines the acidic or alkalinity of nutrient solution | | Seeed Libelium Turtle Tough Atlas Scientific |
| EC sensor | Grove EC sensor GF Signet 2818 EM 802-EC | Determines the nutrient solute concentration balance | | Grove Georg Fischer Sensorex |
| Temperature ure and Humidity sensor | DHT 11,LM35 AHT 21B & A2315C | Measures the atmospheric temperature and humidity level | | Digikey Texas Instrument AOSONG |

*(Table 1) cont.....*

| | | | | |
|---|---|---|---|---|
| Ion Selective Electrode (ISE)/ Ion Meter | Orion Star A214 pH/ ISE INE- PXSJ- 226 | Measures individualconcentration of cation/anion in nutrient mix | | Thermo Fischer Scientific MRC laboratory Instruments |
| PAR sensor | RS-GH-*- AL | Measures the light intensity | | Vernier RENKE |
| $CO_2$ sensor | EE820 $CO_2$ | Checks the hydroponic environment for $CO_2$ concentration | | Galcon Tech Grow |
| Optical sensor | SPAD 502 Plus Chlorophyll meter | Plant health is monitored using leaf chlorophyll | | Konica Minolta Japan |
| Ultrasonic sensor | V-95 | Measures water level | | Massa Sonic |

## Internet of Hydroponics (IoH): Architecture and Working

Automated sensor operation requires seamless connectivity and specialized hardware installation at an early stage. Due to the advent of 4G/5G technology and cost-effective sensor network, this lacuna will soon be overcome. AI, IoT, and big data analytics help to forecast the quality and quantity of vegetables and fruits, which can make the growers efficient in planning the growth and production according to demand and market options. The digital twin will, in turn, help monitor in different agricultural and farming parameters to maintain nutrient solutions for increasing productivity.

IoH is a structure of distributed and interlinked embedded systems interacting through wireless sensor networks used in a hydroponic farming environment. These wireless sensor network is a vital part of IoH as it gathers information through these sensors categorized by numerous levels of communication and reduced consumption of power. WSN consists of sensor nodes capable of sensing various parameters related to soilless farming such as EC, pH, temperature, humidity, and $Co_2$ level. In the core of the WSN node, the microcontrollers process the reading gathered from various sensors and takes different decisions related to the next fertigation cycle. The emergence of IoH and WSN has led to the intelligent automation of hydroponic farming systems.

There are several ways by which IoH can improve urban farming:

i. Smart hydroponic sensors can collect enormous and accurate data regarding the controlled farming environment; thus, the next fertigation cycle can easily be predicted and regulated to maintain proper crop growth progress.
ii. The production through smart hydroponics can easily be regulated to meet the forecasted food demand to help properly sell produced crops.
iii. The automation of hydroponic farming will reduce the effort, time and labour in the field chores and increases business efficiency.
iv. Waste reduction and cost management goals can be achieved with increased control over hydroponic farming.
v. The production quality and quantity will improve through the automation of hydroponic farming.

Fig. (**6**) presents the IoH architecture for data-driven hydroponic agriculture. The sensor data is stored at the device level and processed at the edge computing layer to derive diagnostic features. In the next layer, *i.e.*, the communication layer, the data is transmitted to different nodes and IoT gateway units. Here, all data is stored in a local database through Ethernet/ WiFi. In the fog computing layer, the data is acquired, processed, and stored in a database. In the cloud computing layer, the locally stored database is transferred to the cloud to develop the application for the end users. The last layer, *i.e.*, the application layer, provides the user with data insights regarding yields, mobile applications and fertigation cycle, *etc.*

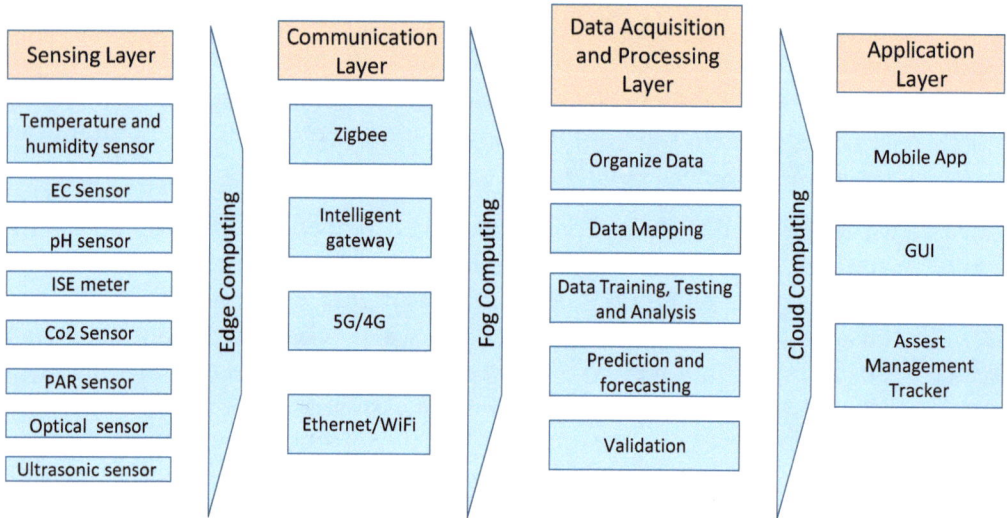

**Fig. (6).** IoH Architecture.

The few related works in the field of IoT and AI-based hydroponic farming are summarized in Table **2**.

**Table 2. Related work in the field of smart hydroponic farming enabled with IoT.**

| References | Architecture | Crops Grown | Sensors | Merits | Demerits |
|---|---|---|---|---|---|
| [10] | IoT, Raspberry Pi, Arduino uno, Solar based | Pakchoy, lettuce, kale | NFT DS18B20 temperature sensor, TDS meter, pH & EC sensor, HC-SR04 ultrasonic sensor | Web interface is provided from remote monitoring control | High initial cost |
| [7] | iHydroIOT | - | Water level, $Co_2$, humidity, light sensor | iOS based mobile application | System is not implemented for real plant growth |
| [11] | Cloud IoT | Potato | Blight MOTE Sensor | Pest prevention for late blight potato | Requires a collection of weather reports |

*(Table 2) cont.....*

| References | Architecture | Crops Grown | Sensors | Merits | Demerits |
|---|---|---|---|---|---|
| [12] | IoT and deep DNN | Tomato | Arduino, Raspberry Pi3, LDR, DHT11 sensor, tensor flow | Intelligent monitoring of plant growth | Expert farmers are required to handle the intelligent farming system |
| [13] | IoT and Node MCU Blynk | - | Node MCU ESP8266, DHT11, pH level monitoring | A cost-effective pH regulating module is developed | Actual implementation on plant growth is not demonstrated |
| [14] | IoT for aero hydroponic farming | potato | AHT15, OPT3001, air quality sensor | Two layers farming can be done | Complex in construction |
| [15] | IoT with blue LED lights | Lettuce | pH probe, UV, humidity, temperature sensors | Blue LED lights help in accumulation of leaf density and pigmentation | Cost addition for supplying the light modules. |
| [16] | SVM with Slack API bot server | Tomato | pH, humidity, water level sensors | Good for vegetable cultivation | Memory of cloud is limited. |

## MACHINE LEARNING AND DATA MINING IN HYDROPONICS

Machine learning (ML), data analytics, cloud computing, and block chain play key roles in ensuring food security and solve many issues related to productivity, water conservation, soil properties, and plant health. There is a lot of data generated in the agricultural supply chain, which enables producers and organizers to draw insights into demand and productivity through data-driven models. However, a data-driven supply chain faces challenges in terms of data collection, storage, visualization, accuracy, and data security.

The ML is categorized into three different learning techniques such as supervised, unsupervised and reinforcement learning. In supervised learning, a predictive model is created using labelled data with prior knowledge of input variables and desired output. ML such as Random forests, Bayesian network, decision trees and regression analysis are grouped in supervised learning. Unsupervised learning establishes hidden patterns between unlabelled datasets and uses dimensionality reduction. ANN, Deep learning and clustering comes under unsupervised learning. In reinforcement learning, the training and testing are combined and the learner interacts with the environment to collect information. Reinforcement learning can be utilized in robot navigation, machine skill acquisition and real-time decision making. Fig. (**7**) depicts the applications of machine learning in the hydroponic supply chain, such as irrigation management, crop selection, disease detection, transport, consumer analysis, *etc.*

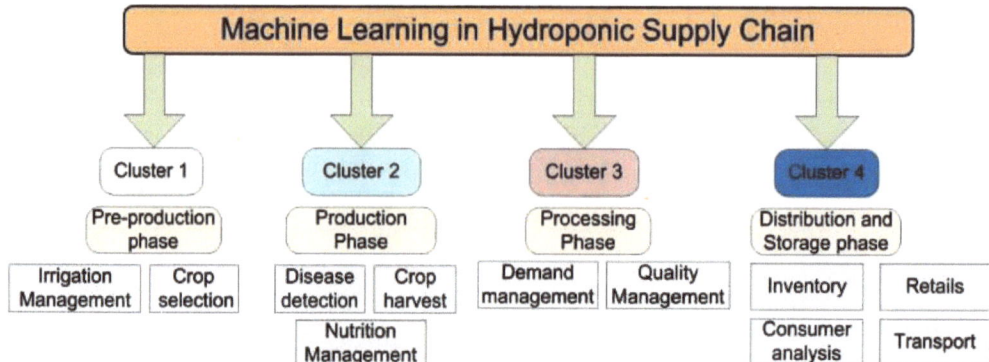

**Fig. (7).** Machine Learning in Hydroponics Supply Chain.

## Applications of Machine Learning

a. **Predicting soil properties and yield forecasting:** Accurate prediction of soil conditions leads to efficient soil management practices. ML is also used for crop selection and maximizes the net yield of crops over the season.

b. **Irrigation management:** ML can be used for efficient irrigation management based on simulation and optimization techniques. It plays a vital role in the quality and quantity of crop production.

c. **Crop disease detection:** Machine learning and pattern recognition can be used for site specific management of diseases and weed detection.

d. **Crop quality management:** These practices help in nutrient management to produce the right market price for crops.

e. **Harvesting:** Remote sensing data and an ensemble learning model are used for predicting coffee yield in the harvest phase. To forecast the yield during harvest, data mining techniques such as mean clustering, ANN, SVM and KNN are implemented.

f. **Production and distribution planning:** An ML algorithm helps in improving demand forecasting, production and distribution planning.

g. **Transportation:** The majority of studies on vehicle routine, minimizing product damage and preserving product quality are accomplished using genetic algorithms. ML algorithm is used for estimating freight, inventory management, estimation of shelf life, and dynamic allocation.

h. **Consumer analytics:** ML techniques such as deep learning and ANN are used in the food retailing phase for predicting consumer perception, demand, and buying behavior.

i. **Inventory management:** It is an essential driver of cost as it utilises capital cost and utilizes storage spaces.

## Challenges Faced in Machine Learning

1. **Data security:** The problems related to the use of data, data storage, data ownership, and data access are dealt with in data security techniques.
2. **Infrastructure:** Internet connectivity is significant for the adoption of technology and automation in hydroponic farming. For this, the infrastructure for the internet network should be developed.
3. **Standardization of data:** The data generated during the sensor network communication need to be smooth and accurate. In order to fulfil this goal, the converted data should be standardized.
4. **Device interoperability:** To ensure successful penetration of ML techniques in hydroponic farming, the device interoperability issues should be resolved prior to the functioning of the farming environment.

Many farmers do not have access to the internet, mobile phones and training on the new technologies and data interpretation. Advisories are required to be developed to assist the farmers in understanding the information and recommend suitable mechanisms to improve farm productivity.

## Use Cases of ML in Hydroponics

a. **For accessing the freshness of hydroponic food crops**: there have been wide research on the technology focused on automation of hydroponic farming, however, the quality of the produced crops can be largely improved by the use of image processing and machine learning. Being able to sort the vegetables based on their freshness and withered status can bring benefits to both consumers and producers. In order to categorize the vegetables (*i.e.*, read oak lettuce) in fresh and withered categories, their roots are wrapped with moist packing and kept in opaque boxes consisting of Raspberry Pi, camera and LED light. These instruments can capture the pictures at an interval of 10 minutes for 7 days. Starting 3 days, images are kept in the fresh category while the rest days are kept in withered categories for data training purpose. These images are processed using the image processing toolbox in MATLAB, and 21 features are extracted to train the machine learning classifier, such as the maximum value of RGB channels individually, Standard deviation of RGB channels, mean value *etc*.
   There are 3 classical and deep neural networks trained and compared using a percentage split scheme due to the time taken to train the deep neural network.

The percentage split considered was 80/20. The three classical machine learning classifiers are Decision Tree (J48), Naive Bayes classifier and Multilayer perceptron. The deep neural network used Inception V3 architecture on the dataset. It is found from the results that the Decision tree outperforms other classifiers in terms of cross-validation and accurate categorization (98.06%) of vegetables. The major lacuna of deep neural networks is the time taken to train the network and adequate dataset requirement to perform better as compared to other classifiers [3].

b. **Prediction of crop diseases using machine learning and data mining:** The timely detection and early diagnosis of plant diseases can be achieved using the Bayesian Network, Artificial Intelligence, and other multicriteria approaches of decision support. Here the data was fed to expert SINTA software and the data was classified in three phases. In the first phase, fungal data are structured based on the sign of disease. Weights are evaluated and assigned to the detected diseases in the second phase. In the last phase, recommended improvements were included to improve the expert system [17]. Naïve Bayes, in addition to Bayesian Networks, can be used for the diagnosis of early rice diseases in crops [18]. Similarly, Segmented Fractal Texture analysis and Local Ternary Patterns can be used to develop a model to predict leaf diseases in potato and corn [19]. There are several tools that can be used for machine learning classifiers in addition to MATLAB, such as Weka, R studio, PySpyder, Knime, and RapidMiner. The major tasks accomplished by these tools are portioning of the dataset into training and testing, parameter optimization, model validation using k –fold validation, or cross validation [20].

## Future Area of Research in Hydroponics

There are certainly a few challenges to establish hydroponics farming as a base urban farming method when it is compared with soiled-based farming. Few major concerns that can be taken up as future areas of research in hydroponics are:

1. Initial installation cost is very high compared to the final produce gained at the end of the harvest period compared to soil-based farming.
2. Required trained or skilled personnel to operate the setup for hydroponic farming. Thus it is hard to say that this technique will gain popularity in rural regions.
3. Only limited varieties of crops could be grown in the hydroponic culture. Moreover, for the growth of specific crops, the required hydroponic farming system should be chosen. A detailed study is of crop requirements are needed to be performed before establishing the hydroponic system.

4. The crop production from hydroponics appears to look healthy and fresh, but the taste has not been able to match the soil-based farming yields.
5. Fault tolerance and reliability of the hydroponic setup (*i.e.*, sensors and communication network) are also important, as any change in the plant environment can cause a larger effect on the final output.

## CONCLUDING REMARKS

Hydroponic farming is emerging as a non-trivial solution to upcoming issues of food security due to the fast growth in population and an exponential increase in food demand. The automation of the farming environment is only possible with the proper utilization of wired/wireless sensor networks, machine learning and data mining algorithms and communication technologies. There are various advantages of opting for urban farming, such as limited space requirement, efficient use of water, and energy to produce healthy crops in limited time and cost. However, to achieve these goals, skilled technical personnel are required to setup and operate the farming environment. Moreover, the government regulatory policies and state designated agencies should motivate these schemes for larger scale adoption. This chapter discusses the architecture, components, structure, challenges, advantages, and future scope of hydroponic farming.

## REFERENCES

[1]     M.I. Alipio, A.E.M. Dela Cruz, J.D.A. Doria, and R.M.S. Fruto, "On the design of nutrient film technique hydroponics farm for smart agriculture", *Eng. Agric. Environ. Food,* vol. 12, no. 3, pp. 315-324, 2019.
[http://dx.doi.org/10.1016/j.eaef.2019.02.008]

[2]     P. Srivani, and S.H. Manjula, "A controlled environment agriculture with hydroponics: Variants, parameters, methodologies and challenges for smart farming", *2019 Fifteenth International Conference on Information Processing,* Bengaluru, India, pp. 1-8, 2019.
[http://dx.doi.org/10.1109/ICInPro47689.2019.9092043]

[3]     K. Wongpatikaseree, N. Hnoohom, and S. Yuenyong, "Machine learning methods for assessing freshness in hydroponic produce", *2018 International Joint Symposium on Artificial Intelligence and Natural Language Processing,* Pattaya, Thailand, pp. 1-4, 2018.
[http://dx.doi.org/10.1109/iSAI-NLP.2018.8692883]

[4]     D. Adidrana, and N. Surantha, "Hydroponic nutrient control system based on internet of things and k-nearest neighbors", *2019 International Conference on Computer, Control, Informatics and its Applications,* Tangerang, Indonesia, pp. 166-171, 2019.

[5]     S. Gertphol, P. Chulaka, and T. Changmai, "Predictive models for lettuce quality from internet of things-based hydroponic farm", *2018 22nd International Computer Science and Engineering Conference,* Chiang Mai, Thailand, pp. 1-5, 2018.
[http://dx.doi.org/10.1109/ICSEC.2018.8712676]

[6]     S. Adhau, R. Surwase, and K.H. Kowdiki, "Design of fully automated low cost hydroponic system using labview and AVR microcontroller", *2017 IEEE International Conference on Intelligent Techniques in Control, Optimization and Signal Processing (INCOS),* Srivilliputtur, India, pp. 1-4, 2017.
[http://dx.doi.org/10.1109/ITCOSP.2017.8303091]

[7]    G. Marques, D. Aleixo, and R. Pitarma, "Enhanced hydroponic agriculture environmental monitoring: An internet of things approach", In: *International Conference on Computational Science* Springer: Cham, 2019, pp. 658-669.
[http://dx.doi.org/10.1007/978-3-030-22744-9_51]

[8]    T. Namgyel, S. Siyang, C. Khunarak, T. Pobkrut, J. Norbu, T. Chaiyasit, and T. Kerdcharoen, "IoT based hydroponic system with supplementary LED light for smart home farming of lettuce", *2018 15th International Conference on Electrical Engineering/Electronics, Computer, Telecommunications and Information Technology,* Chiang Rai, Thailand, pp. 221-224, 2018.
[http://dx.doi.org/10.1109/ECTICon.2018.8619983]

[9]    S. Tagle, R. Pena, F. Oblea, H. Benoza, N. Ledesma, J. Gonzaga, and L.A.G. Lim, "Development of an automated data acquisition system for hydroponic farming", *2018 IEEE 10th International Conference on Humanoid, Nanotechnology, Information Technology,Communication and Control, Environment and Management,* Baguio City, Philippines, 2018, pp. 1-5, 2018.
[http://dx.doi.org/10.1109/HNICEM.2018.8666373]

[10]   P.N. Crisnapati, I.N.K. Wardana, I.K.A.A. Aryanto, and A. Hermawan, "Hommons: Hydroponic management and monitoring system for an IOT based NFT farm using web technology", *In 2017 5th International Conference on Cyber and IT Service Management,* Denpasar, Indonesia, pp. 1-6, 2017.

[11]   K. Foughali, K. Fathallah, and A. Frihida, "A cloud-IOT based decision support system for potato pest prevention", *Procedia Comput. Sci.,* vol. 160, pp. 616-623, 2019.
[http://dx.doi.org/10.1016/j.procs.2019.11.038]

[12]   M. Mehra, S. Saxena, S. Sankaranarayanan, R.J. Tom, and M. Veeramanikandan, "IoT based hydroponics system using deep neural networks", *Comput. Electron. Agric.,* vol. 155, pp. 473-486, 2018.
[http://dx.doi.org/10.1016/j.compag.2018.10.015]

[13]   S. Pawar, S. Tembe, R. Acharekar, S. Khan, and S. Yadav, "Design of an IoT enabled Automated Hydroponics system using NodeMCU and Blynk", *2019 IEEE 5th International Conference for Convergence in Technology (I2CT),* Bombay, India, pp. 1-6, 2019.
[http://dx.doi.org/10.1109/I2CT45611.2019.9033544]

[14]   H.C. Tang, T.Y. Cheng, J.C. Wong, R.C. Cheung, and A.H. Lam, "Aero- hydroponic agriculture iot system. in 2021 ieee 7th world forum on internet of things", *2021 IEEE 7th World Forum on Internet of Things (WF-IoT),* New Orleans, LA, USA, pp. 741-746, 2021.

[15]   T. Namgyel, S. Siyang, C. Khunarak, T. Pobkrut, J. Norbu, T. Chaiyasit, and T. Kerdcharoen, "IoT based hydroponic system with supplementary LED light for smart home farming of lettuce", *2018 15th International Conference on Electrical Engineering/Electronics, Computer, Telecommunications and Information Technology (ECTI-CON),* Chiang Rai, Thailand, pp. 221-224, 2018.
[http://dx.doi.org/10.1109/ECTICon.2018.8619983]

[16]   N. Kitpo, Y. Kugai, M. Inoue, T. Yokemura, and S. Satomura, "Internet of things for greenhouse monitoring system using deep learning and bot notification services", *2019 IEEE International Conference on Consumer Electronics (ICCE),* Las Vegas, NV, USA, pp. 1-4, 2019.
[http://dx.doi.org/10.1109/ICCE.2019.8661999]

[17]   F. Balducci, D. Impedovo, and G. Pirlo, "Machine learning applications on agricultural datasets for smart farm enhancement", *Machines,* vol. 6, no. 3, p. 38, 2018.
[http://dx.doi.org/10.3390/machines6030038]

[18]   V. Sharma, A.A. Mir, and D.A. Sarwr, "Detection of rice disease using bayes' classifier and minimum distance classifier", *J. Multi. Infor. Sys.,* vol. 7, no. 1, pp. 17-24, 2020.
[http://dx.doi.org/10.33851/JMIS.2020.7.1.17]

[19]   K. Aurangzeb, F. Akmal, M.A. Khan, M. Sharif, and M.Y. Javed, "Advanced machine learning algorithm based system for crops leaf diseases recognition", *2020 6th Conference on Data Science and Machine Learning Applications (CDMA),* Riyadh, Saudi Arabia, pp. 146-151, 2020.

[http://dx.doi.org/10.1109/CDMA47397.2020.00031]

[20]   M. Rashid, B.S. Bari, Y. Yusup, M.A. Kamaruddin, and N. Khan, "A comprehensive review of crop yield prediction using machine learning approaches with special emphasis on palm oil yield prediction", *IEEE Access,* vol. 9, pp. 63406-63439, 2021.
[http://dx.doi.org/10.1109/ACCESS.2021.3075159]

# CHAPTER 4

# Agriculture Robotics

**Bogala Mallikharjuna Reddy**[1,*]

*¹ ACPL, Technology Business Incubator, University of Madras, Guindy Campus, Chennai, India*

**Abstract:** In an agriculture-based society, where sustainable farming operations are required, quantitative field status and plant-by-plant monitoring may benefit all cultivators by enhancing farmland management. Sensing technology, artificial intelligence, autonomous robotics, and computerized data analytics will be important. In this book chapter, the essential features of using robotics in agriculture are presented; namely, the primary reasons for the automation of agriculture, the role of robotics in agriculture, its classification, evolution, and consideration of autonomous navigation for commercial agricultural robots, currently existing models of agriculture robots and their comparison, the potential benefits and limitations of agriculture robotics, gathering of massive data and using data science approaches for improving the food productivity and its influence on boosting the agriculture industry. The current study focuses on the adoption of agriculture robotics in the farming sector for various purposes (from land preparation to harvesting). The application of agriculture robotics for food production can favor the incorporation of agricultural robotics companies to minimize labor costs and food shortages. Furthermore, agriculture robotics can be the catalyst for new sources of information on the environmental impact (agroecological footprint) of the local food production chain.

**Keywords:** Agricultural robotics, Digital agriculture, Smart farming, Data science, Commercial model.

## INTRODUCTION

Agriculture is the foundation of civilization since it primarily provides food and fiber that all humans require to survive [1]. Agriculture is a crucial field for the research and implementation of new technology [2]. The advancement of robotics, science, technology, and engineering has altered the face of the farming sector.

* **Corresponding author Bogala Mallikharjuna Reddy:** ACPL, Technology Business Incubator, University of Madras, Guindy Campus, Chennai, India; Telephone: 091-9000097382; E-mail: mbogala@gmail.com

**S. Gowrishankar, Hamidah Ibrahim, A. Veena, K. P. Asha Rani & A. H. Srinivasa (Eds.)**

## AUTOMATION OF FARMING

Agricultural operations continue to rely significantly on human labor, which is vulnerable to health issues such as the global public health crisis caused by the Corona virus pandemic (COVID-19) that resulted in a huge number of deaths globally (~6,484,136 deaths till 07/09/2022, as per the World Health Organization (WHO) data [4, 5]). This resulted in a variety of restrictions on social and economic activities. In that sense, poor nations that rely significantly on food produced by small farmers, livestock producers, and artisanal fishers would be the worst hit by the pandemic. The Food and Agriculture Organization (FAO) claims that COVID-19's social isolation tactics impede farmers from obtaining input and product marketplaces by raising post-harvest losses [6]. On the other side, because agricultural jobs are considered "hard work" and poor profitability in affluent nations such as the United States, young people are opting for career prospects in urban areas, and therefore farmers are looking for innovative ways to automate their farmland, and to recover any losses [7].

## PRECISION AGRICULTURE ROBOTICS

Fortunately, scientific advances in various areas of human knowledge are transforming the way agricultural activities are managed, increasingly reducing human intervention to address the difficulties imposed by population growth, accelerated urbanization, greater competitiveness of high-quality products, a lack of qualified labor, and the vulnerability of human labor to situations of health risk. Precision agriculture (PA) is described as an agricultural strategy that improves the number of right decisions taken by farmers per unit area of land per unit of time with enhanced net benefits [8]. This concept is more general, allowing technological devices or humans to make decisions. The PA is a management approach that employs electronic information and other techniques to collect, process, and examine spatial and temporal data to direct targeted actions that increase agricultural procedures' efficiency, productivity, and sustainability [9]. This concept clearly defines the application of technology to improve agricultural operations. Precision agriculture robotics (PAR) technologies were classified into three categories: (a) Robotics, (b) Artificial Intelligence (AI), and (c) the Internet of Things (IoT). All these technologies may be used separately or in tandem, as shown in Fig. (**1a**).

The employment of robotics and electronic equipment in different tasks of agriculture, such as land preparation, seeding, planting, pest management, and harvesting, has given a boost to the development of PAR [10]. Precision Agriculture's market worth was predicted to be $3.67 billion in 2016 and expected to rise at a 14.7% annual pace to $7.29 billion in 2021 [11]. Though this book chapter will focus on the utilization of robots in farming, the good technologies of AI and IoT are frequently part of the subcomponents of an application requiring the deployment of robots to perform agricultural tasks. As per Zha, who reviewed the application of AI in farming, AI can be used in weed management, soil management, and collaboration with IoT technologies [12]. The author describes how algorithms for computer vision, such as Convolution Neural Networks (CNN) and Deep Belief Networks (DBN), have shown impressive outcomes in fruit categorization and weed recognition in complex environments, such as those with differing ambient lighting, background complexities, angle of capture, and alteration in the shapes and colors of the weeds/fruits.

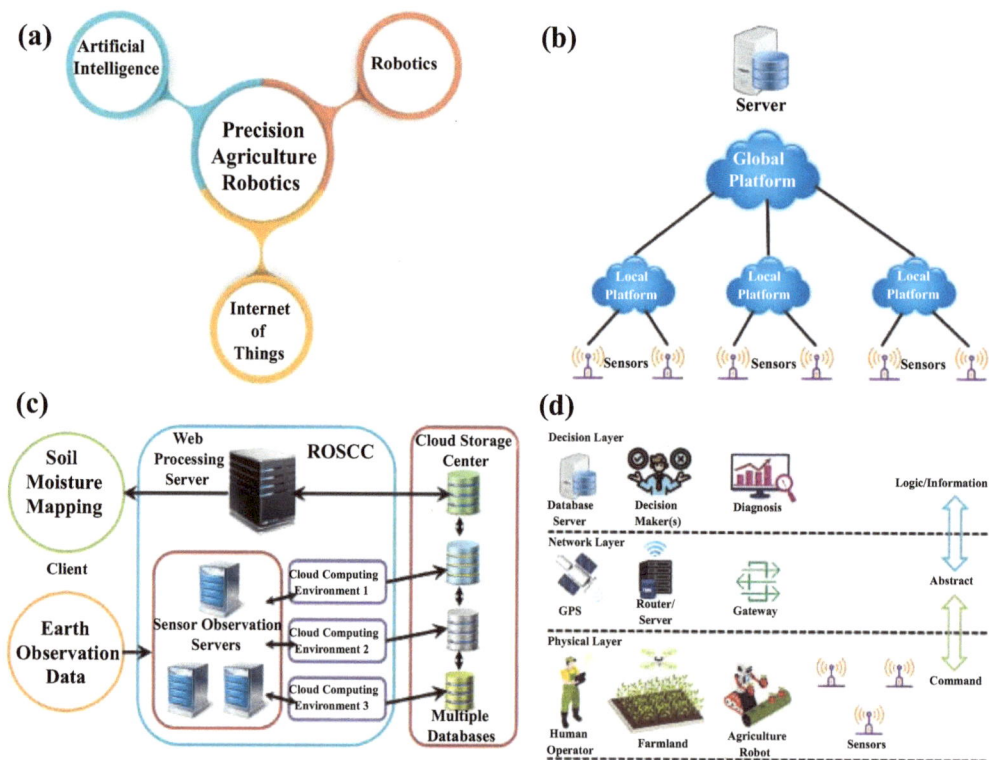

**Fig. (1).** A schematic of (**a**) Precision Agriculture Robotics technological areas, (**b**) IoT structure in agriculture, and architectures of (**c**) ROSCC approach and (**d**) CPS nodes [15].

## IOT-BASED SMART AGRICULTURE

The notion of smart agriculture must be tightly related to the usage of IoT technology, similar to how the idea of smart cities is connected to IoT [4]. Various artificial intelligence approaches, such as CNN, exhibited a high degree of adaptability to quick changes in natural sunlight, changing seasons, and crop growth. Thus, the multiple types of integration of IoT devices being used in agricultural activities, with different kinds of artificial intelligence algorithms, and with the multiple robotic systems reviewed and described in this book chapter may aid in improving process control, monitoring, preservation, and standardization, as well as the development of exact multi-purpose frameworks to fix short-term harvest monitoring and long-term yield estimation issues [13]. Furthermore, because of the ability to exchange Machine-to-Machine (M2M) information, the integration of IoT sensors with mobile robots offers enormous promise for developing the notions of parallelism and swarm of robots [4]. Wang *et al.* investigated the Internet of Things in agricultural architecture with heterogeneous sensor data and presented a data management system integrating cloud computing to allow IoT in agriculture (Fig. **1b**) [14, 15]. Their concept is based on HBase, a two-tier storage structure for a distributed database with high scalability. Their work also presents a cloud-based management method for heterogeneous sensor data for IoT in agriculture. It is made up of a data unification module, an aberrant data processing module, and a two-layer architecture for data storage and access. A cloud computing-based architecture for farm information integration was also developed [16].

Zhou *et al.* tackled the data management challenge of big-size remote sensing pictures in soil moisture mapping for Precision Agriculture [17]. This approach employs a Remote Sensing Observation Sharing approach dependent on cloud computing (ROSCC) to improve remote sensing image storage and perform large-scale mapping of soil moisture using PA (Fig. **1c**) [15]. Cyber-Physical Systems (CPS) must adapt to a dynamic physical reality and constantly extend their capabilities [18]. Three-tier architecture that incorporates fog computing, cloud computing, and actuator and sensor networks were created. The approach employs sensor observation and micro-virtual machines, integrating virtual machine isolation with standardized storage and transmission of information within a Sensor Web Enablement (SWE) framework. The suggested architecture is associated with the Internet of Things and is suited to PA. The surge in digital transformation has created a substantial possibility for more efficient production *via* CPS, which will permit new ideas for future agricultural systems [19]. The fast advancement of information and communication technology is propelling the advancement of mobile machines and gadgets into cyber-physical systems with minimal communication constraints. Nie *et al.* (2014) created a PA architecture based on CPS technology, which includes three control levels: (a) the physical layer, (b) the network layer, and (c) the decision layer (Fig. **1d**) [15, 19].

The current book chapter focuses on the elements that influence agricultural robotics adoption in the contemporary farming industry, the main reasons for agriculture automation, the importance of agriculture robotics, its categorization, evolution, and commercialization of agricultural robots, a comparison of current models, and potential advantages and disadvantages of agriculture robotics. Furthermore, farming robots can serve as catalysts for new sources of data on the agroecological footprint of the food production chain.

## ROBOTICS IN AGRICULTURE

Guo *et al.* developed MDR-CPS, a CPS-oriented framework, and workflow for stress management of agricultural greenhouses [21]. It was created with the goal of monitoring, recognizing, and reacting to various forms of stress. As an integrated CPS, the system incorporates sensors, robots, people, and agricultural greenhouses to monitor, identify, and respond to aberrant circumstances and conditions. The goal is to develop a new system that integrates wireless sensor networks, agricultural robots, and people to identify and respond selectively to pressures as early as feasible using collaborative control theory (CCT). A typical framework of agricultural MDR–CPS is displayed in Fig. (**2**) [15, 20, 21].

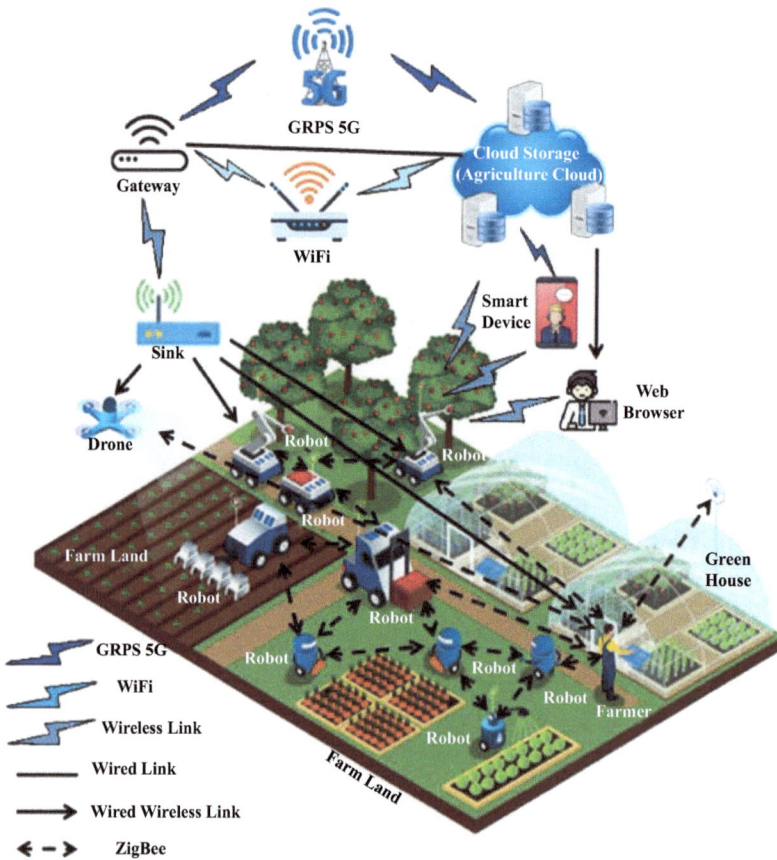

**Fig. (2).** An illustration of the MDR-CPS architecture framework in agriculture [15, 21].

## Classification of Robots

Robots may be classified based on the four major criteria that also identify future robotics markets: (a) operating environment, (b) interaction or cooperation with users, (c) physical format, and (d) function performance (Fig. **3**) [22]. Every robotic system incorporates a diverse set of technologies. Each technology offers particular capabilities that dictate and characterize the overall operation of the system. Robots are equipped with a variety of capacities depending on their usage, such as reconfigurability, motion capabilities, the ability to handle items, sense the surroundings, and act accordingly.

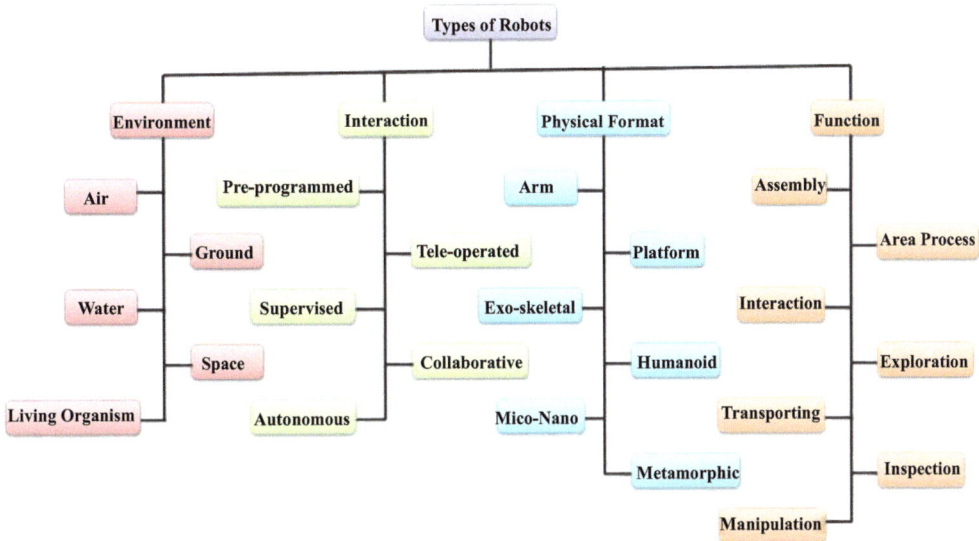

**Fig. (3).** A classification of Robots [22].

## Agriculture Robotics Evolution

Despite considerable improvements in industrial countries throughout the 1st half of the 21st century, the key split in agricultural technology history occurred in the 1950s in the second half of the century and at the beginning of the twenty-first century. The agricultural robotics evolutionary development timeline is displayed in Fig. (**4**) [22]. The world's first touchscreen phone was released in 2000. Although they had fewer capabilities than today's smartphones, they were very popular because they introduced a promising technology for farmers to stay connected with their colleagues, have access to data while on the go, and place orders for agricultural products (*e.g.*, seed or fertilizer) at any place or at any time. Automation technology improves agricultural productivity significantly, but it also faces several technical challenges in terms of resource management and utilization, increased demand for product quality and organic farming products, the complexity of human-machine interactions, data security, mishandling of a large amount of the produced IoT data and the complementarity of labor and technology [23].

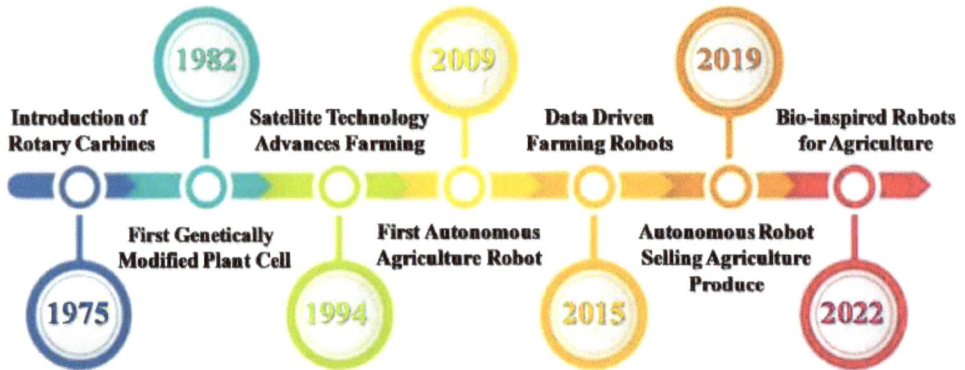

**Fig. (4).** A flowchart showing an evolutionary timeline of agriculture robotics [22].

Information and Communication Technologies (ICT) have an important role in providing major benefits in a variety of fields, such as field information through data analytics and sensing, and autonomous robots are likely to replace manual labor in agricultural tasks (weeding, harvesting, *via*.). Global navigation systems, wireless technologies, management information systems, geographic information systems, monitoring technologies, telematics systems, and superior controlling systems are some examples of ICT. These automation technologies have been successfully implemented in agricultural production. Because they may boost the capacity of the farm robotics system, the operational speed, and the repeatability of various processes and tasks, ICT-supported production systems reduce, to a significant extent, the human needs for decision-making reasoning and sensor technology [24].

**Cooperative Agricultural Robotics**

To minimize the performance gap between installed and realized agricultural machinery, a holistic strategy should be used, taking into account performance at both the local (single machine) and global levels (the whole process). Single-machine capabilities might be improved by integrating systems that offer help or autonomous robotic operations, or overall process performance could be improved by enhancing the interrelationship between single machines [15]. Cooperation can be obtained among agricultural robots in a variety of forms, including (i) unmanned aerial vehicles (UAVs), (ii) human robots, (iii) unmanned ground vehicles (UGVs), (iv) manipulation by multi-arm systems, and (v) hybrid teams of UAVs and UGVs [3]. The designed multi-robot systems incorporate autonomous aerial and field robots with limited operating capabilities to enable the automation of the suggested activities [25]. These are the bare minimum of activities or orders contemplated for both aerial robots and field robots.

Fig. (**5a**) depicts the Mission Manager's general design and connectivity with other modules of the Robot fleets for Highly Effective Agriculture (RHEA) and Forestry Management system [25]. Both aerial and field robots contain sensing devices that permit them to gather information about their internal states and their surroundings. This information is critical for supervising task execution and maintaining control. To execute the inspection, the aerial robots must contain additional sensors (visible and infrared cameras, as well as GPS) to acquire geo-referenced images that can be analyzed later to extract useful information. Ground robots must have precise actuators to complete the task specified. Thus, they differ greatly. Few examples are seeders, sprayers, and harvesting robotic arms [26].

To complete a task on farmland (encompassing several hectares), the team robots must constantly cooperate. Although they do not need to know anything about the other robots in the team, an organization element (Mission Manager) must be part of the multi-robot system; someone must start the robots' operations and deliver them the first plans that they must carry out. As a result, even though there is no direct communication between the robots, the multi-robot system is made aware. The base station computer runs the Mission Manager, which unifies and automates both the inspection and treatment duties. Its primary purpose is to automate the multi-robot system's work sequence. Although the robots are autonomous, high-level software is employed to oversee job execution and combine inspection data with therapy. Supervision can be carried out within the robots themselves or by an external computer that monitors the work of the entire team and is available to the operator. In the distributed multi-level supervision process, each basic supervisor is a component of the Unit Control System (UCS) of each robot (Fig. **5b**) [25]. When defects in the onboard subsystems are discovered and subsystems issue alerts, alarms with identification numbers are created. In some scenarios, errors can be resolved by the subsystem or by the robot's supervisor without the need for operator interaction [25]. Thus, integration and coordination entail the production and transmission of the correct trajectory to every ground robot for completing the treatment, while the supervisor notifies the operator of any unusual behavior seen during job execution.

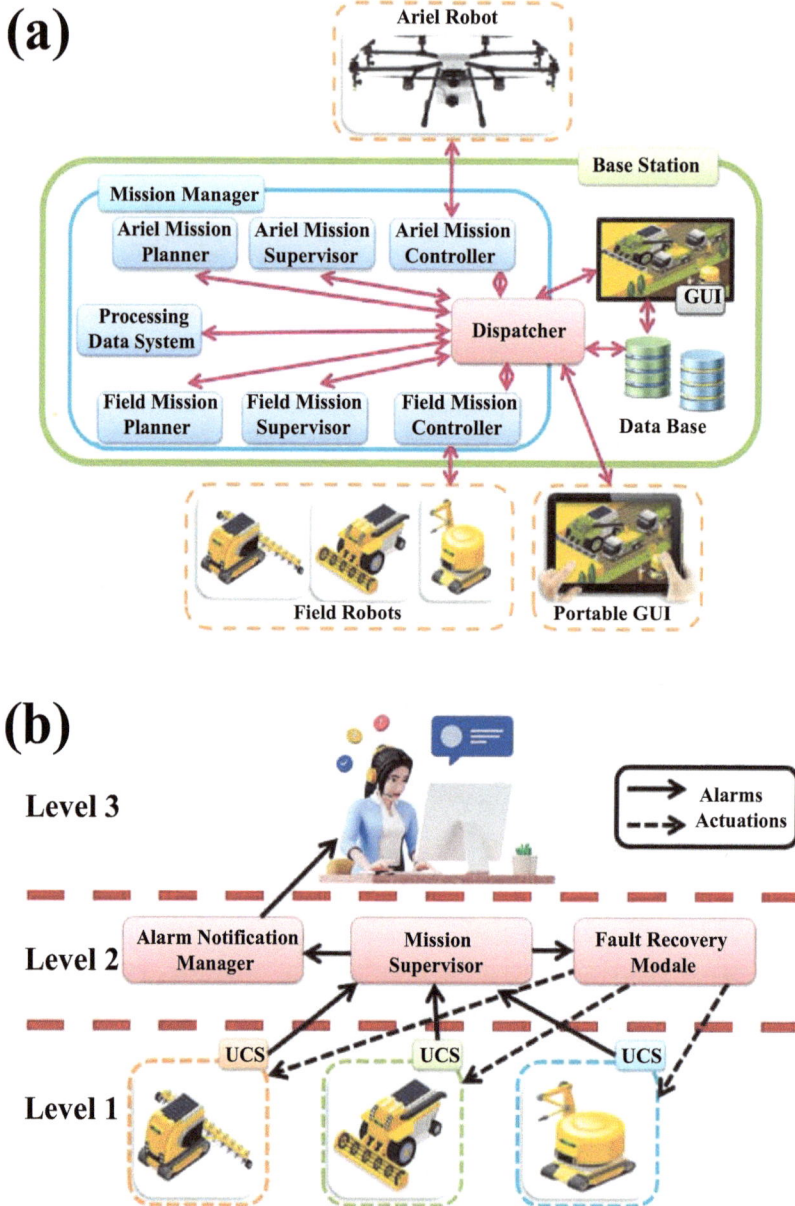

**Fig. (5).** A depiction of (**a**) multi-robot system and (**b**) multi-level supervision of AR [25].

## APPLICATIONS OF AGRICULTURE ROBOTICS

The different activities of agriculture were combined into subdivisions: farmland preparation, sowing, planting, inspection, spraying, treatment, yield estimation, phenotyping, and harvesting [4]. As a result, the following subsections will discuss important farming operations, the importance of various types of agriculture robots and their applications in specific agricultural ecosystems, as well as their commercial aspects.

## Farmland Preparation

One of the primary agricultural works is to prepare the soil for planting, which includes plowing the land and adding fertilizers. Plowing the land (inversion of soil layers) provides for more oxygen entrance and carbon dioxide outflow; yet, depending on local climate circumstances, it might impair future crops by significantly lowering soil carbon reserves. Soil fertilization, on the other hand, replaces the nutrients required for crop development. Fig. (**6**) depicts some of the agricultural robots designed for farmland preparation [4]. Some of these carefully controlled farm robots can work in tough terrain, plow the field, fertilize the soil, and even identify things ahead and come to a halt in an emergency.

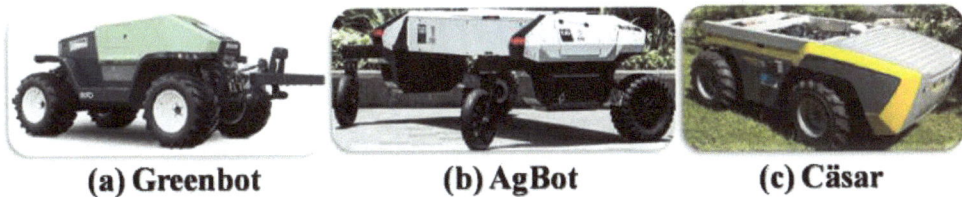

**(a) Greenbot**          **(b) AgBot**          **(c) Cäsar**

**Fig. (6).** Examples of agriculture robots for land preparation prior to planting [4].

## Sowing and Planting

Sowing and planting is the procedure of planting seeds or young plants into the soil for plant growth. This procedure necessitates greater accuracy since various plants necessitate a certain space between each plant to promote development while also maximizing production [1]. The traditional planting process necessitates a farmer physically inserting each seed into the soil. This method takes a lot of time and energy since this procedure demands a lot of consistency and precision, and it generally covers a large agricultural area. Later, specialized planting equipment linked behind the tractor performs sowing and planting chores [4]. Hence, a planter machine has been invented, in which the farmer operates the machine by directing the machine's motion while also planting the seed into the soil.

The designed planter is normally pulled behind a tractor, where it plants the seed in a repetitive motion [1]. Because the tractor and planter are operated by humans, the row's uniformity will suffer because the row will not be in a straight line and there will be some locations where the planter is incapable or fails to plant the seed. Tractors, on the other hand, are heavy machinery, and their frequent movement across the farm increases soil compaction [4]. Soil compaction activity has several negative effects on agricultural environments, including increasing apparent density and soil resistance, decreasing porosity, the rate of water infiltration and aeration, and affecting chemical properties and biogeochemical cycles, which influence plant development and soil biodiversity [27]. Thus, an efficient autonomous system is required that will assure a straight-line crop row and won't miss any seed planting. Small and less expensive robots can be employed on small farms, as opposed to tractors (heavy) that are used on large-scale farms.

Some examples of farm robots used for sowing and planting are illustrated in Fig. (**7**) [1, 4]. Autonomous robotics systems have been developed for a variety of crops, including maize, wheat, sugarcane, and vegetables [28 - 31]. The Agribot platform was used to create an autonomous seeding robot [32]. An Infrared (IR) sensor was utilized in this development to verify the integrity of the seed tank as well as for row identification, and it produced a good result in terms of seed distance accuracy. To ensure planting uniformity, a seed metering unit guidance system has been devised [30, 33]. Another method for ensuring consistency in spacing between seeds during the planting operations is to use a global positioning system. A seed and fertilizer control system has been created employing integrated sensing technologies from the Global Navigation Satellite System (GNSS) and the Inertial Measurement Unit (IMU) [1]. The vehicle will execute the operation in the designed system based on the target application rate of seed and fertilizer by matching the target parameter to the time and space information obtained from IMU/GNSS.

(a) Di-Wheel          (b) Fendt Mars          (c) Tom Robot

**Fig. (7).** Examples of agriculture robots for sowing and planting [1, 4].

Variable-rate seeding control system (VSCS) for planters of maize crop is based on Global Positioning System (GPS) [34]. The receiver (GPS) is employed in this application to determine the real-time seeding position, travel speed, and course angle at a preset frequency. The GPS data is then analyzed to determine latitude, longitude, travel speed, and course angle. A seed spacing uniformity employing a high-speed camera has been created in addition to the global positioning system technology for an effective seed planting operation [35]. A Fuji camera (F660EXR) with a continuous frame rate (320 frames per second) is utilized in this application to track the trajectory of seed dropping, which is mounted at the unissem pneumatic planter output. To assess the sowing rate of wheat, alfalfa, and chickpea seeds in planters, an infrared sensing system comprised of infrared light emitting diodes (IR LEDs) and photodiodes was created [36]. The development of an efficient planter with seed detection will be able to provide an optimum planting procedure with minimal operational costs while ensuring seed distribution consistency throughout the field. Therefore, with better planting capability, automation of the planting process would be very effective and simple to use by the farmers in the coming years.

## Inspection

Following the completion of the sowing stage, the farmer must constantly monitor the planting growth to ensure that it is disease and pest free. In agriculture, inspection is the process of inspecting or observing plants for illnesses or quality flaws. Plant diseases are the primary source of productivity losses in agriculture, resulting in economic losses. According to FAO figures, pests and illnesses destroy 20% - 40% of total agricultural productivity in the world [37]. As the agricultural environment is dynamic, several abnormal and uncertain stress conditions, such as abnormalities in humidity, temperature, water levels, pests, and disease emergence, have affected the plants and their products. Due to the very dynamic nature of the agricultural environment, a variety of unexpected and atypical stress events, such as changes in humidity, temperature, water levels, pests and diseases, have impacted the plants and the agriculture productivity. Weed invasion severely compromises agricultural development and can even destroy crops. Weeds may attract pests as well as small animals such as snakes and mice. As a result, the earlier it is removed, the fewer the financial damages. If those anomalies are not addressed promptly, significant and irreversible harm may result [38]. In Australia, for example, weed management costs close to $4 billion each year [39].

Traditionally, farmers will personally observe plant irregularities utilizing their human visual system to carry out the examination. Farmers' ages have been growing in recent years, reducing the efficiency of inspection operations since the quality of the human visual system depletes with age [1]. Furthermore, the deployment of automation in agricultural inspection necessitates the advancement of the system to substitute the capability of human vision to perform the inspection process more efficiently. As a result, computer vision algorithms have been widely employed in robotics to substitute human vision in agricultural plant inspection. The search for automation of the disease diagnosis process in plants and weed detection is not new; a study in this field has been ongoing since 1998 [4]. A computer vision algorithm is a sophisticated image processing technology that has a positive effect and can substitute human vision in completing certain specific work in the inspection process [40]. A computer vision algorithm is widely utilized in a variety of areas, including agriculture. Image processing and computer vision applications in agriculture have risen as a result of lower equipment prices, improved computational capacity, and rising interest in non-destructive food evaluation methods [41].

The majority of vision system deployments in agriculture are used for disease detection, with others being utilized for product quality monitoring. The inspection process is an essential agricultural activity whose primary goal is to identify plant diseases and preserve the quality of agricultural products. The automation of the inspection process is critical because it reduces the inefficiency of manual inspection procedures used by farmers. Fig. (**8**) depicts some of the vision system implementations in the agricultural inspection process. The action of autonomous inspection is often carried out by mounting a camera in a fixed location on a mobile robot platform or a drone. In addition to using image processing methods in agricultural inspection, some researchers employ hyperspectral imaging for plant disease diagnosis.

**(a) Naïo TED**      **(b) RB-VOGUI**      **(c) AgriBot**

**Fig. (8).** Examples of agriculture robots for inspection of plants [1].

Peanut leaf spots are detected using novel Normalized Differential Spectral Indices (NDSI) [42]. To track the chemical ingredient in agricultural goods, this approach employs an ultrasensitive gray-imaging-based immunochromatography detection method for quantitative analysis. With the growth of the Industrial Revolution 4.0 (IR 4.0), the cyber-physical system has become the primary emphasis area, with all devices and sensors networked in the cloud. As a result, agricultural inspection technology is progressing toward IR 4.0. IoT has recently become a trend in agricultural inspection processes, allowing farmers to monitor and control their agricultural control systems *via* computer and mobile apps [43, 44]. Most IoT applications in agricultural inspection are utilized to track plant diseases in real-time by capturing photos and gathering sensory data from the farm, such as temperature, humidity, moisture, and soil pH level, and displaying it on a website or mobile app. As a result, illnesses such as early blight, light blight, and powdery mildew may be spotted by the farmer before the epidemic spreads. In addition to monitoring farm conditions, IoT technology permits farmers to operate other agricultural systems such as irrigation and spraying, and it can also send warning messages to farmers if there are any irregularities in plant conditions using artificial intelligence [44]. Disease prevention and quality checking of agricultural goods will become more precise and efficient in the future as a result of the autonomous IoT technology design and its deployment in the inspection process *via* smart robots.

## Spraying and Plant Treatment

Spraying and plant treatment are frequently performed by spraying herbicides and pesticides (insecticides and fungicides) after a thorough assessment of diseased plants in the crop [4]. In agriculture, spraying is a common way of administering pest-control chemicals, fertilizers, or growing media to plants as a fine mist for disease treatment and plant growth management. Pest-control chemicals are often administered consistently throughout the fields in most farming operations to control disease transmission. Although numerous pests and plant diseases have an uneven geographical distribution, especially during the early phases of development, this strategy is used [45]. Therefore, to reduce the expense of pest-control chemicals used in agricultural operations, selective spraying was developed and studied during the last two decades [46]. This automated selective spraying method, which is often carried out by completely automated mobile robots or equipment, allows for the selective administration of pesticides just where and when they are required. The primary goal of this selected operation is to limit pesticide consumption while also preventing illness and its wide spread across the greenhouse [47].

Previous research has focused on building an effective spraying system with low operational costs in the improvement of autonomous selective spraying methods. To attain the above goal, a variable rate spraying system that allows farmers to automatically adapt the herbicide or pesticide volume rate to the target depending on the canopy size and treatment required is devised [48]. Fig. (**9**) shows some examples of agriculture robots employed for spraying and plant treatment. In addition to minimizing pesticide usage, several studies are concentrating on navigation management to lower the operational costs of the robot, such as time and energy, while maintaining excellent location tracking. This study area is critical to ensure that the robot can precisely navigate toward the target while incurring the least amount of trip cost during the spraying process. Even though researchers are focusing on lowering operational costs in the design of spraying navigation systems, they are also focusing on maintaining high position accuracy for agricultural robots by using Real-Time Kinematic GPS (RTK-GPS) and Inertial Measurement Unit (IMU) navigation systems equipped with highly accurate position and attitude sensors [1].

(a) Oz      (b) Dino      (c) LadyBird

(d) BoniRob   (e) Robotic sprayer   (f) Pollinator Robot   (g) Swagbot

**Fig. (9).** Examples of agriculture robots for spraying and plant treatment [4].

Although most autonomous spraying operations in agriculture employ a land-based vehicle, some implementations propose the use of an unmanned aerial vehicle (UAV) to perform this task [1]. It is critical to substitute human labor in spraying operations to limit the danger of toxic chemicals to farmers. With the rapid advancement of automation technology, the transition state of human substitution in spraying operations will become considerably easier, with the primary goal of treating plant diseases at the lowest possible operating cost [1]. However, a significant study is still being undertaken to assure that the utilization of automation in spraying operations would bring unlimited advantages to farmers in managing plant diseases and ensuring future food security. As a result, researchers may continue to enhance current methods and strategies for producing an effective spraying operation for agricultural applications.

## Yield and Phenotype Estimating

Farmers may manage their crops more efficiently with reliable data on the quantity and quality of fruit growth offered by increasingly advanced instruments. Yield estimation is monitoring of the entire crop and estimating the amount of fruit produced [4]. Plant development, on the other hand, might be hampered by factors such as climate change and soil quality. Thus, by determining the phenotype of the plants, the genotype may be linked, allowing the optimal development circumstances to be identified. However, it should be highlighted that for a robot to estimate phenotyping and/or yield, it must have both trustworthy sensory data and effective computer vision algorithms [4]. Researchers proposed using sensors and machine vision algorithms to recognize crop rows and gather field data as early as 1998 and 2001 [49, 50]. Plant phenotyping may be retrieved in a variety of ways, including measuring plant height, weight, biomass, shape, colors, volume, light absorption, and temperature [51]. All these factors play a very key role in the management of farming operations for enhanced agriculture productivity. In many situations, cooperation between humans and robots are critical in effective implementation of robotic technology in agriculture. Fig. (**10**) depicts several agricultural robots (EcoRobtix, Hexapod, Vinobot, Shrimp, Phenocopter and AgRob v16) for the estimation of yield and phenotyping. Therefore, agriculture robots come in many ways depending on the intended type of tasks to be performed in agriculture operations and can vary from one crop to another crop.

## Harvesting

Harvesting is the process of gathering agricultural goods to be processed or sold. To run this procedure, the fruits or vegetables must be gathered and kept for future processing or sold straight to purchasers. This method is characterized as a time-consuming and labor-intensive process since it requires extensive observation with repetitive procedures. As a result, the development of autonomous harvesting systems has been active during the previous decades. Several crop implementations have occurred in recent years, including strawberry, apple, tomato, kiwi, capsicum, grape, litchi, citrus, pumpkin, and heavyweight crops [52 - 61]. The majority of implementations are aimed at improving the accuracy of harvesting systems by proposing various approaches and methods with varying software and hardware structural design. Fig. (**11**) provides some examples of agriculture robots used for the harvesting process.

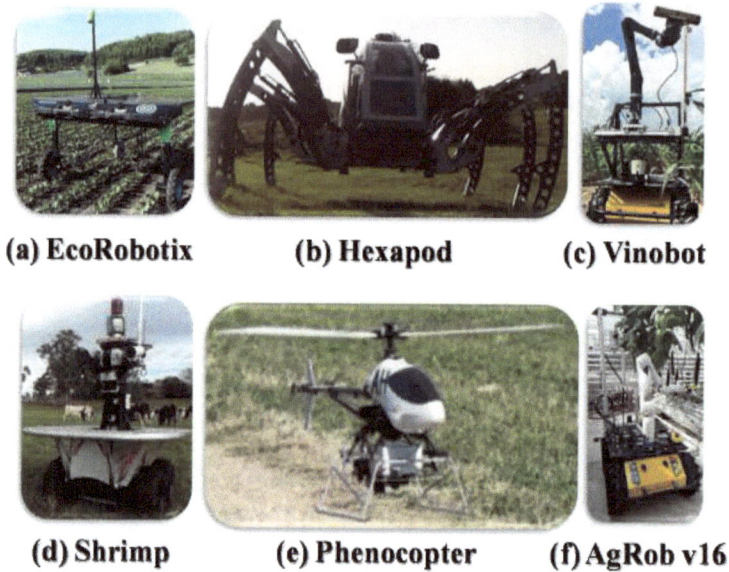

(a) EcoRobotix          (b) Hexapod          (c) Vinobot

(d) Shrimp          (e) Phenocopter          (f) AgRob v16

**Fig. (10).** Examples of agriculture robots for yield estimation and phenotyping [4].

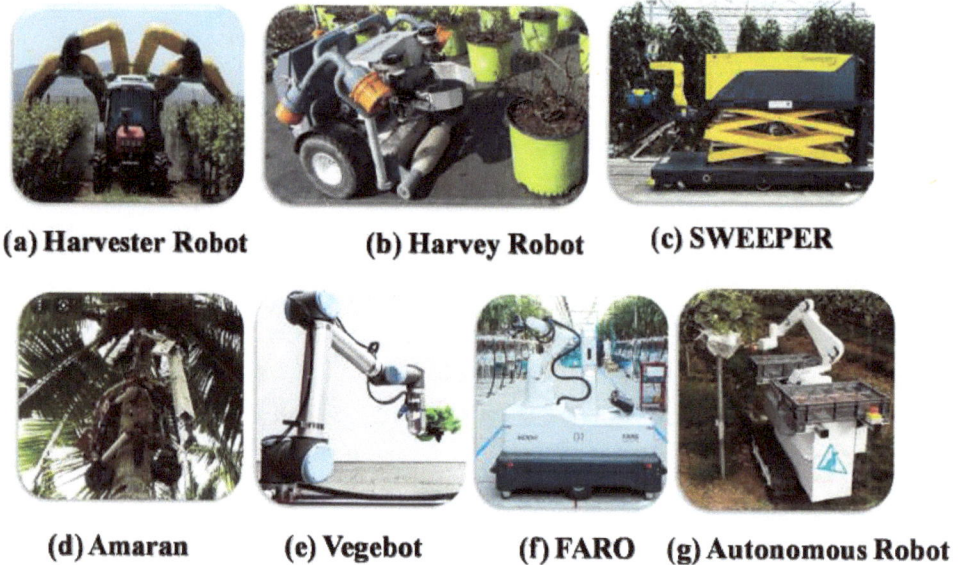

(a) Harvester Robot          (b) Harvey Robot          (c) SWEEPER

(d) Amaran          (e) Vegebot          (f) FARO          (g) Autonomous Robot

**Fig. (11).** Examples of agriculture robots for harvesting crops [4].

Several stages are necessary to carry out an autonomous harvesting procedure. First, the mobile robot must be able to identify the object or area that has to be collected by locating the target location. The robotic arm will then be carefully directed towards the desired area while avoiding any obstructions. Finally, the cutting mechanism will be activated, with the procedure typically beginning with fruit gripping, stem cutting, and the harvested product being placed in a storage compartment incorporated inside the mobile robot structure [1]. As a result, each stage in the autonomous robotic harvesting process represents a separate set of issues that agricultural experts must improve and overcome to produce an effective harvesting robot. The activity of harvesting in agriculture is extremely significant since the quality of the harvested product will also be impacted by the manner the harvesting procedure is being handled owing to its delicate nature [1]. Even if the plants are well-treated during the growth period, the quality of the output cannot be assured since the product may be destroyed during harvesting due to the use of robotics and automation. As a result, several continuing studies are still being undertaken to verify that the efficiency of robotic and automation applications in the harvesting process is comparable to or better than the way humans work to harvest agricultural products and how promptly the agriculture robots perform without compromising their quality in improving the food productivity. Fig. (**12**) illustrates the concept of agriculture robotics (production to harvesting and handling) within the entire agriculture chain system (also includes transportation and processing stages). Table **1** gives the list of different agriculture robots, various sensors used, computer vision algorithms, main purpose, and ultimate applications.

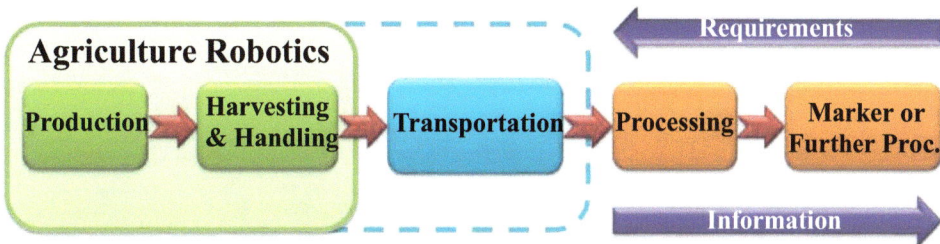

**Fig. (12).**  The concept of agriculture robotics within the agriculture chain system [22].

**Table 1. List of agriculture robots, their sensors, computer algorithms, purpose and applications [4].**

| Agriculture Robot | Sensors Used | Computer Algorithm | Purpose Type | Application Crop | Refs. |
|---|---|---|---|---|---|
| AgBot | RGB camera, compass, accelerometer and RTK GPS | - | Land preparation | Corn | [62] |

*(Table 1) cont.....*

| Agriculture Robot | Sensors Used | Computer Algorithm | Purpose Type | Application Crop | Refs. |
|---|---|---|---|---|---|
| AgBot II | RGB camera | LBP | Mechanical Weeding | Cotton, sow thistle, feather top rhodes grass and wild oats | [63] |
| AGRAS MG-1P | RTK GPS, RGB camera, gyroscope, accelerometer and compass, Omnidirectional radar | - | Land preparation | Rice, soy and corn | [64] |
| Agrob V16 | RTK/GPS/INS, LiDAR, Stereo, RGB-D and RGB cameras | hLBP and SVM | Yield estimation | Grape | [65] |
| Amaran | RGB camera | - | Harvesting | Coconut | [66] |
| Ara ecoRobotix | RGB camera RTK/GPS and compass | - | Phenotyping | General | [67] |
| BoniRob | RGB, NIR cameras and ultrasonic sensor | CNN | Chemical weeding | Sugar beet | [68] |
| Cäsar | RTK GNSS, Ultrasonic | - | Land preparation. | Orchard or vineyard | [69] |
| Disease robot | RGB camera and laser | PCA and CV | Disease Identification | Tomato | [70] |
| Di-Wheel | Smartphone embedded sensors | - | Sowing and planting | Horticulture | [71] |
| eAGROBOT | RGB camera | K-means and Neural Networks | Disease Identification | Cotton and groundnut | [72] |
| GARotics | RGB-D camera | RANSAC and euclidean clustering | Harvesting | Green asparagus | [73] |
| Greenbot | RTK GPS, Bump | - | Land preparation | Horticulture, fruit and arable farming | [74] |
| Harvester robot 1 | RGB-D and ToF cameras | SVM | Harvesting | Aubergines | [75] |
| Harvester robot 2 | RGB-D camera, encoder and LiDAR | HSV color-thresholding | Harvesting | Strawberry | [76] |
| Harvester robot 3 | RGB-D camera | Dasnet, 3D-SHT and Octree | Harvesting | Apple | [77] |

*(Table 1) cont.....*

| Agriculture Robot | Sensors Used | Computer Algorithm | Purpose Type | Application Crop | Refs. |
|---|---|---|---|---|---|
| Harvester robot 4 | RGB and laser sensors | R-YOLO | Harvesting | Strawberry | [78] |
| Harvey plataform | RGB-D camera, pressure and separation sensors | DCNN | Harvesting | Sweet pepper | [79] |
| Hexapod | $CO_2$ gas module, anemoscope and infrared distance sensor | - | Yield estimation | General | [80] |
| Kubota farm vehicle | LiDAR, GPS and IMU | Continuous-Time SLAM | Yield estimation | Grape | [81] |
| K-Weedbot | RGB camera | Hough transform | Mechanical Weeding | Paddy | [82] |
| LadyBird | RTK/GPS/INS, Hyperspectral and thermal cameras and LiDAR | ExG-ExR | Chemical weeding | Lettuce, cauliflower and broccoli | [83] |
| Lumai-5 | Angle and speed | - | Sowing and planting | Wheat | [84] |
| Pheno-Copter | RGB and thermal cameras and LiDAR | RANSAC and DEM | Phenotyping | Sorghum, sugarcane and wheat | [85] |
| Pollinator robot | RGB camera and Odometry | CNN | Pollination | Kiwi | [86] |
| Pruning robot 1 | RGB camera | SVM | Pruning | Grape | [87] |
| SAVSAR | RGB camera and LiDAR | FDA and GDA | Chemical weeding | Grape | [88] |
| Shrimp | RGB camera | MLP and CNN | Yield estimation | Apple | [89] |
| Sowing robot | Ultrasonic | - | Sowing and planting | Corn | [90] |
| Swagbot | GPS, RGB-D, IR and hiperspectral cameras and LiDAR | NDVI | General | General | [91] |
| SWEEPER | RGB-D camera | Deep learning, shape, color-based detection and HT | Harvesting | Sweet pepper | [92] |
| TerraSentia | RGB camera, RTK/GPS and LiDAR | LiDAR-based navigation | Phenotyping | Corn | [93] |
| Vegebot | RGB camera | R-CNN | Harvesting | Lettuce | [94] |

*(Table 1) cont.....*

| Agriculture Robot | Sensors Used | Computer Algorithm | Purpose Type | Application Crop | Refs. |
|---|---|---|---|---|---|
| VINBOT | RTK, DGPS, LiDAR, RGB and NIR cameras | NDVI | Yield estimation | Grape | [95] |
| VineRobot | FA-Sense LEAF, FA-Sense ANTH, ultrasonic and RGB camera | Chlorophyll-based fluorescence and RGB machine vision | Yield estimation | Grape | [96] |
| Vinobot | Stereo camera, environmental Sensors, DGPS and LiDAR | VisualSFM | Phenotyping | Corn | [51] |
| Vinoculer | Stereo RGB and IR cameras and air temperature sensors | VisualSFM | Phenotyping | Corn | [97] |
| Weeding robot 1 | RGB-D camera | RANSAC | Mechanical weeding | Broccoli and lettuce | [98] |

## Commercial Agricultural Robots

To accomplish specified tasks, Commercial Agricultural Robots (CARs) must assess their surroundings, find themselves, and travel around a crop field. Sensors and navigation techniques are two of the most important parts of building autonomous unmanned ground vehicles [99]. Sensor and cost requirements vary substantially depending on the sort of environment (unstructured vs. structured) in which the robot will work. As a result, the sensors used will have a considerable influence on the unit's ultimate cost as well as the robot's navigation ability. Automated guided vehicles (AGVs) are successful agriculture robots in structured situations [100]. Autonomous or semi-autonomous trolleys, for example, carry items inside warehouses at automobile plants. These carts are monitored by AGV technology and run on supporting infrastructure in a predetermined map for self-localization. Supportive infrastructure entails the installation of basic magnetic tape tracks on the floor, cables beneath the floor transmitting a magnetic signature, or laser-identified landmarks throughout the production [99]. In organized situations, these navigation strategies improve the vehicle's autonomy and safety while using fewer sensors. This type of technology can be employed in agriculture for particular jobs in greenhouses, however, it is not immediately adaptable to outdoor and unstructured conditions.

CARs, on the other hand, may operate in both organized and unstructured situations. They commonly employ a Global Navigation Satellite System (GNSS) for field localization [99]. However, due to the signal deficiency in some areas and under the canopy, it is preferable to have systems that do not rely solely on GNSS to provide long-term autonomy. Non-GNSS-based localization methods like SLAM (Simultaneous Location And Mapping) are effective in a variety of applications [101, 102]. Therefore, strong localization in the field necessitates the deployment of many approaches at the same time. The choice of a navigation technique to be employed in an unstructured environment is a critical difficulty in the advancement of CARs. Because the needed skills of self-localization and navigation vary greatly depending on the field and application, the robot's implemented design can influence the number and kind of the CARs sensors. Fig. (**13**) shows a few examples of CARs available in the market for the farming sector.

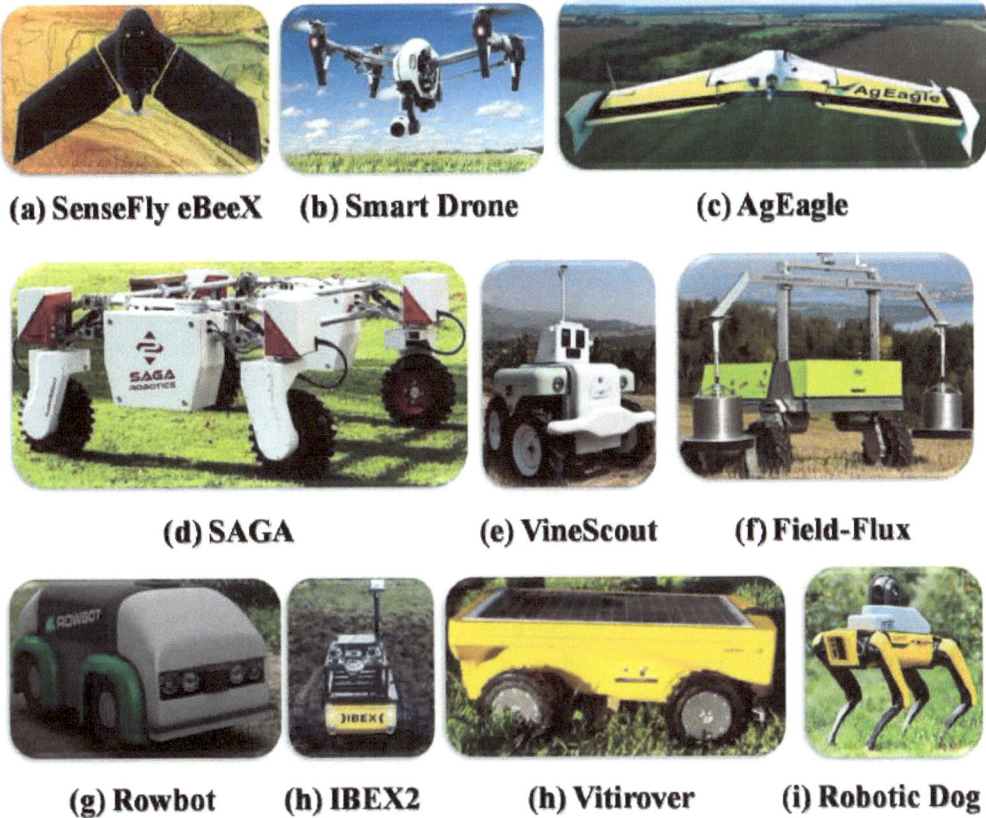

(a) SenseFly eBeeX    (b) Smart Drone        (c) AgEagle

(d) SAGA        (e) VineScout    (f) Field-Flux

(g) Rowbot    (h) IBEX2    (h) Vitirover    (i) Robotic Dog

**Fig. (13).** Examples of some commercial agriculture robots [99].

## CHALLENGES IN AGRICULTURE ROBOTICS

Many study fields in cooperative agricultural robots also are open to further advancement to enhance present systems, both in terms of utility and dependability, and therefore attain commercial availability in the coming years. In terms of human-robot collaboration teams, it is critical to investigate acceptable interfaces to provide successful collaborative control, particularly for field employees with limited technological knowledge. Furthermore, human-robot cooperation remains a very promising study topic that will equip robots with a greater sense of human actions and intentions, substantially improving coordination concerns. Monitoring duties, on the other hand, necessitate the deployment of a vast number of cooperative aerial and ground vehicles. Despite the increasing availability of low-cost small-scale robotic platforms, it appears that the power needs of such robots limit the number of jobs they can accomplish, limiting them to brief inspection and mapping missions. Finally, the cooperative implementation of robots and humans in the management of agricultural products, as used in harvesting and shipping activities, is a potential study area in the near future.

In addition to becoming necessary for specific tasks, such as occlusion removal or handling, the employment of numerous arms in a cooperative way may minimize the limits of other underlying technologies, such as vision. Table **2** highlights some advantages and disadvantages of cooperative techniques used in agricultural robotics [3]. Distinct CAR operations have different issues, necessitating a variety of solutions to address the individual operating issue. To reduce system mistakes during future deployment, the improvement process of effective autonomous robotics in agriculture must examine all possibilities and obstacles in all types of agricultural operations. Furthermore, the development costs should be accurately recognized to guarantee that farmers may spend their wealth as consumers. As a result, it is quite likely that autonomous robotics in agriculture will be widely accepted around the world in the future.

Table 2. A comparision of advantages and disadvantages of agriculture robotics [3].

| Agriculture Robotics | |
|---|---|
| **Advantages** | **Disadvantages** |
| Decrease of task period. | Constraints in power self-sufficiency. |
| Finishing of tasks where heterogeneous robots are needed. | Complexity in the human-robot interface and action cooperation is mandatory. |
| Minimizing soil compaction *via* the utilization of smaller-size robots. | Lack of compatibility between humans and multiple robots. |
| Decrease in failure cases of the hardware.. | Human-robot work safety is a threat. |

*(Table 2) cont.....*

| Agriculture Robotics | |
|---|---|
| **Advantages** | **Disadvantages** |
| Can replace any labor scarcity of contact workers. | - |
| Improved accessibility of low-cost robotic tools and platforms. | - |
| - | - |

## CONCLUSION AND FUTURE OUTLOOK

Agriculture robotics and automation play a critical role in ensuring global food security. With the extensive technology provided by the established system, the adoption of robots has enabled farmers to conduct agricultural tasks in a timely way. As the development of robotic systems in agriculture is generally focusing on imitating the behavior of human labor in the completion of agricultural operations, complex multi-tasks such as planting, inspection, spraying, and harvesting will be carried out efficiently with minimal operational costs and human labor agriculture robots. As every agricultural operation requires different features and specifications depending on the unique ecosystem and plant species type, several studies are still being undertaken to verify that the produced autonomous system is more efficient and error-free. Various issues remain for agricultural robotics experts to tackle, and several extensive studies are being done to study the causes and answers to the challenges. Systematic autonomous robotics in agriculture may be designed in the future by combining all of the technologies developed for each operation to create a robust and efficient agricultural robotic system to be widely implemented by farmers all over the world with the main goal of producing a high amount of agricultural output to maintain food security in the future.

## REFERENCES

[1]     M.S.A. Mahmud, M.S.Z. Abidin, A.A. Emmanuel, and H.S. Hasan, "Robotics and automation in agriculture: Present and future applications", *Applications of Modelling and Simulation,* vol. 4, no. 0, pp. 130-140, 2020.

[2]     R. Sparrow, and M. Howard, "Robots in agriculture: Prospects, impacts, ethics, and policy", *Precis. Agric.,* vol. 22, no. 3, pp. 818-833, 2021.
[http://dx.doi.org/10.1007/s11119-020-09757-9]

[3]     C. Lytridis, V.G. Kaburlasos, T. Pachidis, M. Manios, E. Vrochidou, T. Kalampokas, and S. Chatzistamatis, "An overview of cooperative robotics in agriculture", *Agronomy,* vol. 11, no. 9, p. 1818, 2021.
[http://dx.doi.org/10.3390/agronomy11091818]

[4]     L.F.P. Oliveira, A.P. Moreira, and M.F. Silva, "Advances in agriculture robotics: A state-of-the-art review and challenges ahead", *Robotics,* vol. 10, no. 2, p. 52, 2021.
[http://dx.doi.org/10.3390/robotics10020052]

[5]     I. Bergeri, H.C. Lewis, L. Subissi, A. Nardone, M. Valenciano, B. Cheng, K. Glonti, B. Williams, I.O.O. Abejirinde, A. Simniceanu, A. Cassini, R. Grant, A. Rodriguez, A. Vicari, L. Al Ariqi, T. Azim, P.R. Wijesinghe, S.C. Rajatonirina, J.C. Okeibunor, L.V. Le, M. Katz, A. Vaughan, P. Jorgensen, G. Freidl, R. Pebody, and M.D. Van Kerkhove, "Early epidemiological investigations: World Health Organization UNITY protocols provide a standardized and timely international investigation framework during the COVID-19 pandemic", *Influenza Other Respir. Viruses,* vol. 16, no. 1, pp. 7-13, 2022.
[http://dx.doi.org/10.1111/irv.12915] [PMID: 34611986]

[6]     M.A. Mekouar, "Food and agriculture organization of the united nations (FAO)", *Yearbook of International Environmental Law,* vol. 29, pp. 448-468, 2018.
[http://dx.doi.org/10.1093/yiel/yvz057]

[7]     D.V. Rosero, F. Soto Mas, R. Sebastian, S. Guldan, V. Casanova, and L. Nervi, "COVID-19 prevalence and prevention behaviors among us certified organic producers", *J. Occup. Environ. Med.,* vol. 63, no. 12, pp. e937-e943, 2021.
[http://dx.doi.org/10.1097/JOM.0000000000002411] [PMID: 34623976]

[8]     A. McBratney, B. Whelan, T. Ancev, and J. Bouma, "Future directions of precision agriculture", *Precis. Agric.,* vol. 6, no. 1, pp. 7-23, 2005.
[http://dx.doi.org/10.1007/s11119-005-0681-8]

[9]     J. Lowenberg-DeBoer, and B. Erickson, "Setting the record straight on precision agriculture adoption", *Agron. J.,* vol. 111, no. 4, pp. 1552-1569, 2019.
[http://dx.doi.org/10.2134/agronj2018.12.0779]

[10]    J. Lowenberg-DeBoer, I.Y. Huang, V. Grigoriadis, and S. Blackmore, "Economics of robots and automation in field crop production", *Precis. Agric.,* vol. 21, no. 2, pp. 278-299, 2020.
[http://dx.doi.org/10.1007/s11119-019-09667-5]

[11]    L.G. Santesteban, "Precision viticulture and advanced analytics. A short review", *Food Chem.,* vol. 279, pp. 58-62, 2019.
[http://dx.doi.org/10.1016/j.foodchem.2018.11.140] [PMID: 30611512]

[12]    J. Zha, "Artificial Intelligence in Agriculture", *J. Phys. Conf. Ser.,* vol. 1693, no. 1, p. 012058, 2020.
[http://dx.doi.org/10.1088/1742-6596/1693/1/012058]

[13]    F. Cui, "Deployment and integration of smart sensors with IoT devices detecting fire disasters in huge forest environment", *Comput. Commun.,* vol. 150, pp. 818-827, 2020.
[http://dx.doi.org/10.1016/j.comcom.2019.11.051]

[14]    H.Z. Wang, G.W. Lin, J.Q. Wang, W.L. Gao, Y.F. Chen, and Q.L. Duan, "Management of big data in the internet of things in agriculture based on cloud computing", *Appl. Mech. Mater.,* vol. 548-549, pp. 1438-1444, 2014.
[http://dx.doi.org/10.4028/www.scientific.net/AMM.548-549.1438]

[15]    A.S. Nair, S.Y. Nof, and A. Bechar, *"Emerging Directions of Precision Agriculture and Agricultural Robotics,"* in *Innovation in Agricultural Robotics for Precision Agriculture: A Roadmap for Integrating Robots in Precision Agriculture.,* A. Bechar, Ed., Springer International Publishing: Cham, 2021, pp. 177-210.
[http://dx.doi.org/10.1007/978-3-030-77036-5_8]

[16]    Y. Duan, *Design of agriculture information integration and sharing platform based on cloud computing.,* 2012.
[http://dx.doi.org/10.1109/CYBER.2012.6319958]

[17]    L. Zhou, N. Chen, Z. Chen, and C. Xing, "ROSCC: An efficient remote sensing observation-sharing method based on cloud computing for soil moisture mapping in precision Agriculture", *IEEE J. Sel. Top. Appl. Earth Obs. Remote Sens.,* vol. 9, no. 12, pp. 5588-5598, 2016.
[http://dx.doi.org/10.1109/JSTARS.2016.2574810]

[18]   J.V. Pradilla, and C.E. Palau, *Micro Virtual Machines (MicroVMs) for Cloud-assisted Cyber-Physical Systems (CPS)," in Internet of Things.,* R. Buyya, A. Vahid Dastjerdi, Eds., Morgan Kaufmann, 2016, pp. 125-142.
[http://dx.doi.org/10.1016/B978-0-12-805395-9.00007-1]

[19]   T. Herlitzius, "Automation and robotics: The trend towards cyber physical systems in agriculture business", *SAE Technical Paper, Warrendale, PA, SAE Technical Paper,* 2017.
[http://dx.doi.org/10.4271/2017-01-1932]

[20]   J. Nie, R.Z. Sun, and X.H. Li, "A precision agriculture architecture with cyber-physical systems design technology", *Appl. Mech. Mater.,* vol. 543-547, pp. 1567-1570, 2014.
[http://dx.doi.org/10.4028/www.scientific.net/AMM.543-547.1567]

[21]   P. Guo, P.O. Dusadeerungsikul, and S.Y. Nof, "Agricultural cyber physical system collaboration for greenhouse stress management", *Comput. Electron. Agric.,* vol. 150, pp. 439-454, 2018.
[http://dx.doi.org/10.1016/j.compag.2018.05.022]

[22]   D. Bochtis, and S. Moustakidis, *"Mobile Robots: Current Advances and Future Perspectives," in Innovation in Agricultural Robotics for Precision Agriculture: A Roadmap for Integrating Robots in Precision Agriculture.,* A. Bechar, Ed., Springer International Publishing: Cham, 2021, pp. 1-15.
[http://dx.doi.org/10.1007/978-3-030-77036-5_1]

[23]   V. Marinoudi, C.G. Sørensen, S. Pearson, and D. Bochtis, "Robotics and labour in agriculture. A context consideration", *Biosyst. Eng.,* vol. 184, pp. 111-121, 2019.
[http://dx.doi.org/10.1016/j.biosystemseng.2019.06.013]

[24]   C.G. Sørensen, L. Pesonen, D.D. Bochtis, S.G. Vougioukas, and P. Suomi, "Functional requirements for a future farm management information system", *Comput. Electron. Agric.,* vol. 76, no. 2, pp. 266-276, 2011.
[http://dx.doi.org/10.1016/j.compag.2011.02.005]

[25]   A. Ribeiro, and J. Conesa-Muñoz, *"Multi-robot Systems for Precision Agriculture," in Innovation in Agricultural Robotics for Precision Agriculture: A Roadmap for Integrating Robots in Precision Agriculture.,* A. Bechar, Ed., Springer International Publishing: Cham, 2021, pp. 151-175.
[http://dx.doi.org/10.1007/978-3-030-77036-5_7]

[26]   S. Blackmore, "Robotic agriculture: The future of agricultural mechanisation", *Euro. Conf. on Prec. Agri.,* pp. 621-628.

[27]   M.F. Nawaz, G. Bourrié, and F. Trolard, "Soil compaction impact and modelling. A review", *Agron. Sustain. Dev.,* vol. 33, no. 2, pp. 291-309, 2013.
[http://dx.doi.org/10.1007/s13593-011-0071-8]

[28]   A. Cay, H. Kocabiyik, and S. May, "Development of an electro-mechanic control system for seed-metering unit of single seed corn planters Part II: Field performance", *Comput. Electron. Agric.,* vol. 145, pp. 11-17, 2018.
[http://dx.doi.org/10.1016/j.compag.2017.12.021]

[29]   L. Haibo, D. Shuliang, L. Zunmin, and Y. Chuijie, "Study and experiment on a wheat precision seeding robot", *J. Robot,* vol. 2015, p. 12, 2015.
[http://dx.doi.org/10.1155/2015/696301]

[30]   K. Saengprachatanarug, C. Chaloemthoi, K. Kamwilaisak, P. Kasemsiri, S. Chaun-Udom, and E. Taira, "Effect of metering device arrangement to discharge consistency of sugarcane billet planter", *Eng. Agric. Environ. Food,* vol. 11, no. 3, pp. 139-144, 2018.
[http://dx.doi.org/10.1016/j.eaef.2018.03.002]

[31]   Z.M. Khazimov, G.C. Bora, K.M. Khazimov, M.Z. Khazimov, I.B. Ultanova, and A.K. Niyazbayev, "Development of a dual action planting and mulching machine for vegetable seedlings", *Eng. Agric. Environ. Food,* vol. 11, no. 2, pp. 74-78, 2018.
[http://dx.doi.org/10.1016/j.eaef.2018.02.003]

[32] N.S. Naik, V.S Virendra, and R.D. Shruti, "Precision agriculture robot for seeding function", *2016 International Conference on Inventive Computation Technologies.* Coimbatore, India, 2016, pp. 1-3, 2016.
[http://dx.doi.org/10.1109/INVENTIVE.2016.7824880]

[33] X. He, T. Cui, D. Zhang, J. Wei, M. Wang, Y. Yu, Q. Liu, B. Yan, D. Zhao, and L. Yang, "Development of an electric-driven control system for a precision planter based on a closed-loop PID algorithm", *Comput. Electron. Agric.,* vol. 136, pp. 184-192, 2017.
[http://dx.doi.org/10.1016/j.compag.2017.01.028]

[34] W. Fu, N. Gao, X. An, and J. Zhang, "Study on precision application rate technology for maize no-tillage planter in north china plain", *IFAC-PapersOnLine,* vol. 51, no. 17, pp. 412-417, 2018.
[http://dx.doi.org/10.1016/j.ifacol.2018.08.186]

[35] Z. Abdolahzare, and S. Abdanan Mehdizadeh, "Real time laboratory and field monitoring of the effect of the operational parameters on seed falling speed and trajectory of pneumatic planter", *Comput. Electron. Agric.,* vol. 145, pp. 187-198, 2018.
[http://dx.doi.org/10.1016/j.compag.2018.01.001]

[36] B. Besharati, H. Navid, H. Karimi, H. Behfar, and I. Eskandari, "Development of an infrared seed-sensing system to estimate flow rates based on physical properties of seeds", *Comput. Electron. Agric.,* vol. 162, pp. 874-881, 2019.
[http://dx.doi.org/10.1016/j.compag.2019.05.041]

[37] R. Akhter, and S.A. Sofi, "Precision agriculture using IoT data analytics and machine learning", *Journal of King Saud University - Computer and Information Sciences,* vol. 34, no. 8, pp. 5602-5618, 2022.
[http://dx.doi.org/10.1016/j.jksuci.2021.05.013]

[38] A.A. Abba Ari, A. Gueroui, N. Labraoui, and B.O. Yenke, "Concepts and evolution of research in the field of wireless sensor networks", *International journal of Computer Networks & Communications,* vol. 7, no. 1, pp. 81-98, 2015.
[http://dx.doi.org/10.5121/ijcnc.2015.7106]

[39] J.A. Sinden, and G. Griffith, "Combining economic and ecological arguments to value the environmental gains from control of 35 weeds in Australia", *Ecol. Econ.,* vol. 61, no. 2-3, pp. 396-408, 2007.
[http://dx.doi.org/10.1016/j.ecolecon.2006.03.011]

[40] M.K. Tripathi, and D.D. Maktedar, "A role of computer vision in fruits and vegetables among various horticulture products of agriculture fields: A survey", *Inf. Process. Agric.,* vol. 7, no. 2, pp. 183-203, 2020.
[http://dx.doi.org/10.1016/j.inpa.2019.07.003]

[41] S. Mahajan, A. Das, and H.K. Sardana, "Image acquisition techniques for assessment of legume quality", *Trends Food Sci. Technol.,* vol. 42, no. 2, pp. 116-133, 2015.
[http://dx.doi.org/10.1016/j.tifs.2015.01.001]

[42] T. Chen, J. Zhang, Y. Chen, S. Wan, and L. Zhang, "Detection of peanut leaf spots disease using canopy hyperspectral reflectance", *Comput. Electron. Agric.,* vol. 156, pp. 677-683, 2019.
[http://dx.doi.org/10.1016/j.compag.2018.12.036]

[43] E. Suganya, S. Sountharrajan, S.K. Shandilya, and M. Karthiga, *IoT in Agriculture Investigation on Plant Diseases and Nutrient Level Using Image Analysis Techniques,"* in *Internet of Things in Biomedical Engineering.,* V.E. Balas, L.H. Son, S. Jha, M. Khari, R. Kumar, Eds., Academic Press, 2019, pp. 117-130.
[http://dx.doi.org/10.1016/B978-0-12-817356-5.00007-3]

[44] A. Khattab, S.E.D. Habib, H. Ismail, S. Zayan, Y. Fahmy, and M.M. Khairy, "An IoT-based cognitive monitoring system for early plant disease forecast", *Comput. Electron. Agric.,* vol. 166, p. 105028, 2019.

[http://dx.doi.org/10.1016/j.compag.2019.105028]

[45]   R. Oberti, M. Marchi, P. Tirelli, A. Calcante, M. Iriti, E. Tona, M. Hočevar, J. Baur, J. Pfaff, C. Schütz, and H. Ulbrich, "Selective spraying of grapevines for disease control using a modular agricultural robot", *Biosyst. Eng.,* vol. 146, pp. 203-215, 2016.
       [http://dx.doi.org/10.1016/j.biosystemseng.2015.12.004]

[46]   M.E.R. Paice, P.C.H. Miller, and W. Day, "Control requirements for spatially selective herbicide sprayers", *Comput. Electron. Agric.,* vol. 14, no. 2-3, pp. 163-177, 1996.
       [http://dx.doi.org/10.1016/0168-1699(95)00046-1]

[47]   R. Oberti, "CROPS Agricultural Robot: Application to Selective Spraying of Grapevine's Diseases", Available From: https://www.google.com/search?client=firefox-b-d&q=CROPS+Agricultural+Robot%3A+Application+to+Selective+Spraying+of+Grapevine%E2%80%99s+Diseases (Accessed on Sep 01 2022).

[48]   H. Maghsoudi, S. Minaei, B. Ghobadian, and H. Masoudi, "Ultrasonic sensing of pistachio canopy for low-volume precision spraying", *Comput. Electron. Agric.,* vol. 112, pp. 149-160, 2015.
       [http://dx.doi.org/10.1016/j.compag.2014.12.015]

[49]   N. Noguchi, J.F. Reid, E.R. Benson, and T.S. Stombaugh, "Vision intelligence for an agricultural mobile robot using a neural network", *IFAC Proceedings,* pp. 139-144, 1998.
       [http://dx.doi.org/10.1016/S1474-6670(17)42112-4]

[50]   N. Noguchi, J.F. Reid, K. Ishii, and H. Terao, "Multi-Spectrum image sensor for detecting crop status by robot tractor", *IFAC Proceedings,* vol. 34, no. 19, pp. 111-115, 2001.
       [http://dx.doi.org/10.1016/S1474-6670(17)33122-1]

[51]   A. Shafiekhani, S. Kadam, F. Fritschi, and G. DeSouza, "Vinobot and vinoculer: Two robotic platforms for high-throughput field phenotyping", *Sensors,* vol. 17, no. 12, p. 214, 2017.
       [http://dx.doi.org/10.3390/s17010214] [PMID: 28124976]

[52]   Y. Xiong, C. Peng, L. Grimstad, P.J. From, and V. Isler, "Development and field evaluation of a strawberry harvesting robot with a cable-driven gripper", *Comput. Electron. Agric.,* vol. 157, pp. 392-402, 2019.
       [http://dx.doi.org/10.1016/j.compag.2019.01.009]

[53]   W. Ji, Z. Qian, B. Xu, Y. Tao, D. Zhao, and S. Ding, "Apple tree branch segmentation from images with small gray-level difference for agricultural harvesting robot", *Optik,* vol. 127, no. 23, pp. 11173-11182, 2016.
       [http://dx.doi.org/10.1016/j.ijleo.2016.09.044]

[54]   Z. Li, F. Miao, Z. Yang, and H. Wang, "An anthropometric study for the anthropomorphic design of tomato-harvesting robots", *Comput. Electron. Agric.,* vol. 163, p. 104881, 2019.
       [http://dx.doi.org/10.1016/j.compag.2019.104881]

[55]   L. Mu, G. Cui, Y. Liu, Y. Cui, L. Fu, and Y. Gejima, "Design and simulation of an integrated end-effector for picking kiwifruit by robot", *Inf. Process. Agric.,* vol. 7, no. 1, pp. 58-71, 2020.
       [http://dx.doi.org/10.1016/j.inpa.2019.05.004]

[56]   T. Moon, J. Park, and J.E. Son, "Prediction of the fruit development stage of sweet pepper (Capsicum annum var. annuum) by an ensemble model of convolutional and multilayer perceptron", *Biosyst. Eng.,* vol. 210, pp. 171-180, 2021.
       [http://dx.doi.org/10.1016/j.biosystemseng.2021.08.017]

[57]   L. Luo, Y. Tang, Q. Lu, X. Chen, P. Zhang, and X. Zou, "A vision methodology for harvesting robot to detect cutting points on peduncles of double overlapping grape clusters in a vineyard", *Comput. Ind.,* vol. 99, pp. 130-139, 2018.
       [http://dx.doi.org/10.1016/j.compind.2018.03.017]

[58]   J. Zhuang, C. Hou, Y. Tang, Y. He, Q. Guo, Z. Zhong, and S. Luo, "Computer vision-based localisation of picking points for automatic litchi harvesting applications towards natural scenarios",

*Biosyst. Eng.,* vol. 187, pp. 1-20, 2019.
[http://dx.doi.org/10.1016/j.biosystemseng.2019.08.016]

[59]  Y. Wang, Y. Yang, C. Yang, H. Zhao, G. Chen, Z. Zhang, S. Fu, M. Zhang, and H. Xu, "End-effector with a bite mode for harvesting citrus fruit in random stalk orientation environment", *Comput. Electron. Agric.,* vol. 157, pp. 454-470, 2019.
[http://dx.doi.org/10.1016/j.compag.2019.01.015]

[60]  A. Roshanianfard, and N. Noguchi, "Characterization of pumpkin for a harvesting robot", *IFAC-PapersOnLine,* vol. 51, no. 17, pp. 23-30, 2018.
[http://dx.doi.org/10.1016/j.ifacol.2018.08.056]

[61]  T. Kamata, A. Roshanianfard, and N. Noguchi, "Heavy-weight crop harvesting robot: Controlling algorithm", *IFAC-PapersOnLine,* vol. 51, no. 17, pp. 244-249, 2018.
[http://dx.doi.org/10.1016/j.ifacol.2018.08.165]

[62]  N. Khan, G. Medlock, S. Graves, and S. Anwar, "GPS guided autonomous navigation of a small agricultural robot with automated fertilizing system", In: *SAE Technical Paper* PA, SAE Technical Paper: Warrendale, 2018.
[http://dx.doi.org/10.4271/2018-01-0031]

[63]  C.S. McCool, J. Beattie, J. Firn, C. Lehnert, J. Kulk, O. Bawden, R. Russell, and T. Perez, "Efficacy of mechanical weeding tools: A study into alternative weed management strategies enabled by robotics", *IEEE Robot. Autom. Lett.,* pp. 1-1, 2018.
[http://dx.doi.org/10.1109/LRA.2018.2794619]

[64]  D.E. Fedorov, "Modern trends in developing robotic systems in agro-industrial complex", *IOP Conf. Ser. Earth Environ. Sci.,* vol. 949, no. 1, p. 012016, 2022.
[http://dx.doi.org/10.1088/1755-1315/949/1/012016]

[65]  R. Reis, "Redundant robot localization system based in wireless sensor network", *2018 IEEE International Conference on Autonomous Robot Systems and Competitions* 25-27 April 2018, Torres Vedras, Portugal, pp. 154-159.
[http://dx.doi.org/10.1109/ICARSC.2018.8374176]

[66]  R.K. Megalingam, S. Kuttankulangara Manoharan, S.M. Mohan, S.R.R. Vadivel, R. Gangireddy, S. Ghanta, S. Kotte, S.T. Perugupally, and V. Sivanantham, "Amaran: An unmanned robotic coconut tree climber and harvester", *IEEE/ASME Trans. Mechatron.,* pp. 1-1, 2020.
[http://dx.doi.org/10.1109/TMECH.2020.3014293]

[67]  S. Gorjian, and S. Minaei, "Applications of solar PV systems in agricultural automation and robotics", *Photovoltaic Solar Energy Conversion,* pp. 191-235, 2020.
[http://dx.doi.org/10.1016/B978-0-12-819610-6.00007-7]

[68]  X. Wu, S. Aravecchia, P. Lottes, C. Stachniss, and C. Pradalier, "Robotic weed control using automated weed and crop classification", *J. Field Robot.,* vol. 37, no. 2, pp. 322-340, 2020.
[http://dx.doi.org/10.1002/rob.21938]

[69]  L. Emmi, and P. Gonzalez-de-Santos, "Mobile robotics in arable lands: Current state and future trends", *2017 European Conference on Mobile Robots (ECMR),* Paris, France, pp. 1-6, 2017.
[http://dx.doi.org/10.1109/ECMR.2017.8098694]

[70]  N. Schor, A. Bechar, T. Ignat, A. Dombrovsky, Y. Elad, and S. Berman, "Robotic disease detection in greenhouses: Combined detection of powdery mildew and tomato spotted wilt virus", *IEEE Robot. Autom. Lett.,* vol. 1, no. 1, pp. 354-360, 2016.
[http://dx.doi.org/10.1109/LRA.2016.2518214]

[71]  S.K. Samantaray, and S.S. Rout, "Design and development of a di-wheel multipurpose robot for smart agriculture application", In: *in Smart and Sustainable Technologies: Rural and Tribal Development Using IoT and Cloud Computing,* Singapore, 2022, pp. 373-379.
[http://dx.doi.org/10.1007/978-981-19-2277-0_35]

[72]    S.K. Pilli, B. Nallathambi, S.J. George, and V. Diwanji, "eAGROBOT &#x2014; A robot for early crop disease detection using image processing", *2015 2nd International Conference on Electronics and Communication Systems* Coimbatore, India, Feb. 2015, pp. 1684–1689 [http://dx.doi.org/10.1109/ECS.2015.7124873]

[73]    A. Leu, M. Razavi, L. Langstadtler, D. Ristic-Durrant, H. Raffel, C. Schenck, A. Graser, and B. Kuhfuss, "Robotic green asparagus selective harvesting", *IEEE/ASME Trans. Mechatron.,* vol. 22, no. 6, pp. 2401-2410, 2017. [http://dx.doi.org/10.1109/TMECH.2017.2735861]

[74]    J. Spencer, "Is this the tractor of the future?: Machinery & implements", *Farmer's Weekly,* vol. 2016, no. 16013, pp. 50-51, 2016. [http://dx.doi.org/10.10520/EJC186740]

[75]    D. Sepulveda, R. Fernandez, E. Navas, M. Armada, and P. Gonzalez-De-Santos, "Robotic aubergine harvesting using dual-arm manipulation", *IEEE Access,* vol. 8, pp. 121889-121904, 2020. [http://dx.doi.org/10.1109/ACCESS.2020.3006919]

[76]    Y. Xiong, Y. Ge, L. Grimstad, and P.J. From, "An autonomous strawberry harvesting robot: Design, development, integration, and field evaluation", *J. Field Robot.,* vol. 37, no. 2, pp. 202-224, 2020. [http://dx.doi.org/10.1002/rob.21889]

[77]    H. Kang, H. Zhou, and C. Chen, "Visual perception and modeling for autonomous apple harvesting", *IEEE Access,* vol. 8, pp. 62151-62163, 2020. [http://dx.doi.org/10.1109/ACCESS.2020.2984556]

[78]    Y. Yu, K. Zhang, H. Liu, L. Yang, and D. Zhang, "Real-time visual localization of the picking points for a ridge-planting strawberry harvesting robot", *IEEE Access,* vol. 8, pp. 116556-116568, 2020. [http://dx.doi.org/10.1109/ACCESS.2020.3003034]

[79]    C. Lehnert, C. McCool, I. Sa, and T. Perez, "Performance improvements of a sweet pepper harvesting robot in protected cropping environments", *J Field Robotics,* p. 21973, 2020. [http://dx.doi.org/10.1002/rob.21973]

[80]    M. Iida, D. Kang, M. Taniwaki, M. Tanaka, and M. Umeda, "Localization of $CO_2$ source by a hexapod robot equipped with an anemoscope and a gas sensor", *Comput. Electron. Agric.,* vol. 63, no. 1, pp. 73-80, 2008. [http://dx.doi.org/10.1016/j.compag.2008.01.016]

[81]    T. Lowe, P. Moghadam, E. Edwards, and J. Williams, "Canopy density estimation in perennial horticulture crops using 3D spinning lidar SLAM", *J. Field Robot.,* vol. 38, no. 4, pp. 598-618, 2021. [http://dx.doi.org/10.1002/rob.22006]

[82]    K.H. Choi, S.K. Han, S.H. Han, K.H. Park, K.S. Kim, and S. Kim, "Morphology-based guidance line extraction for an autonomous weeding robot in paddy fields", *Comput. Electron. Agric.,* vol. 113, pp. 266-274, 2015. [http://dx.doi.org/10.1016/j.compag.2015.02.014]

[83]    K.F. Francis, P. Colignon, P. Hastir, E. Haubruge, and C. Gaspar, "Evolution of aphidophagous ladybird populations in a vegetable crop and implications as biological agents", *Mededelingen van de Faculteit Landbouwkundige en Toegepaste Biologische Wetenschappen,* 2001. Available From:https://orbi.uliege.be/handle/2268/34188 (Accessed on Sep 08 2022).

[84]    H.B. Lin, C.J. Yi, and Z.M. Liu, "Experimental study on quadruped wheel robot for wheat precision seeding", *Key Eng. Mater.,* vol. 693, pp. 1651-1657, 2016. [http://dx.doi.org/10.4028/www.scientific.net/KEM.693.1651]

[85]    S. Chapman, T. Merz, A. Chan, P. Jackway, S. Hrabar, M. Dreccer, E. Holland, B. Zheng, T. Ling, and J. Jimenez-Berni, "Pheno-Copter: A low-altitude, autonomous remote-sensing robotic helicopter for high-throughput field-based phenotyping", *Agronomy,* vol. 4, no. 2, pp. 279-301, 2014. [http://dx.doi.org/10.3390/agronomy4020279]

[86]  H. Williams, M. Nejati, S. Hussein, N. Penhall, J.Y. Lim, M.H. Jones, J. Bell, H.S. Ahn, S. Bradley, P. Schaare, P. Martinsen, M. Alomar, P. Patel, M. Seabright, M. Duke, A. Scarfe, and B. MacDonald, "Autonomous pollination of individual kiwifruit flowers: Toward a robotic kiwifruit pollinator", *J. Field Robot.*, vol. 37, no. 2, pp. 246-262, 2020.
[http://dx.doi.org/10.1002/rob.21861]

[87]  T. Botterill, S. Paulin, R. Green, S. Williams, J. Lin, V. Saxton, S. Mills, X.Q. Chen, and S. Corbett-Davies, "A robot system for pruning grape vines", *J. Field Robot.*, vol. 34, no. 6, pp. 1100-1122, 2017.
[http://dx.doi.org/10.1002/rob.21680]

[88]  G. Adamides, C. Katsanos, I. Constantinou, G. Christou, M. Xenos, T. Hadzilacos, and Y. Edan, "Design and development of a semi-autonomous agricultural vineyard sprayer: Human-robot interaction aspects", *J. Field Robot.*, vol. 34, no. 8, pp. 1407-1426, 2017.
[http://dx.doi.org/10.1002/rob.21721]

[89]  S. Bargoti, and J.P. Underwood, "Image segmentation for fruit detection and yield estimation in apple orchards", *J. Field Robot.*, vol. 34, no. 6, pp. 1039-1060, 2017.
[http://dx.doi.org/10.1002/rob.21699]

[90]  M.U. Hassan, M. Ullah, and J. Iqbal, "Towards autonomy in agriculture: Design and prototyping of a robotic vehicle with seed selector", *2016 2nd International Conference on Robotics and Artificial Intelligence,* Rawalpindi, Pakistan, pp. 37-44, 2016.
[http://dx.doi.org/10.1109/ICRAI.2016.7791225]

[91]  N.D. Wallace, H. Kong, A.J. Hill, and S. Sukkarieh, "Energy aware mission planning for WMRs on uneven terrains", *IFAC-PapersOnLine,* vol. 52, no. 30, pp. 149-154, 2019.
[http://dx.doi.org/10.1016/j.ifacol.2019.12.513]

[92]  B. Arad, J. Balendonck, R. Barth, O. Ben-Shahar, Y. Edan, T. Hellström, J. Hemming, P. Kurtser, O. Ringdahl, T. Tielen, and B. Tuijl, "Development of a sweet pepper harvesting robot", *J. Field Robot.*, vol. 37, no. 6, pp. 1027-1039, 2020.
[http://dx.doi.org/10.1002/rob.21937]

[93]  V.A.H. Higuti, A.E.B. Velasquez, D.V. Magalhaes, M. Becker, and G. Chowdhary, "Under canopy light detection and ranging-based autonomous navigation", *J. Field Robot.*, vol. 36, no. 3, pp. 547-567, 2019.
[http://dx.doi.org/10.1002/rob.21852]

[94]  S. Birrell, J. Hughes, J.Y. Cai, and F. Iida, "A field-tested robotic harvesting system for iceberg lettuce", *J. Field Robot.*, vol. 37, no. 2, pp. 225-245, 2020.
[http://dx.doi.org/10.1002/rob.21888] [PMID: 32194355]

[95]  C.M. Lopes, "Vineyard yeld estimation by VINBOT robot : preliminary results with the white variety Viosinho", In: *Proceedings 11th Int. Terroir Congress.,* G. Jones, N. Doran, Eds., Southern Oregon University: Ashland, USA, 2016, pp. 458-463. Available from: https://www.repository.utl.pt/handle/10400.5/13128

[96]  M.P. Diago, and J. Tardaguila, "A new robot for vineyard monitoring", *Wine Vitic. J,* 2015. vol. 30, no. 3.
[http://dx.doi.org/10.3316/informit.256729208411211]

[97]  A. Shafiekhani, F. B. Fritschi, and G. DeSouza, "Vinobot and vinoculer: From real to simulated platforms", *in Autonomous Air and Ground Sensing Systems for Agricultural Optimization and Phenotyping III,* Orlando: United States, p. 18, 2018.
[http://dx.doi.org/10.1117/12.2316341]

[98]  J. Gai, L. Tang, and B.L. Steward, "Automated crop plant detection based on the fusion of color and depth images for robotic weed control", *J. Field Robot.*, vol. 37, no. 1, pp. 35-52, 2020.
[http://dx.doi.org/10.1002/rob.21897]

[99]  G. Gil, D.E. Casagrande, L.P. Cortés, and R. Verschae, "Why the low adoption of robotics in the

farms? Challenges for the establishment of commercial agricultural robots", *Smart Agricultural Technology,* vol. 3, p. 100069, 2023.
[http://dx.doi.org/10.1016/j.atech.2022.100069]

[100] G.T. Aguiar, G.A. Oliveira, K.H. Tan, N. Kazantsev, and D. Setti, "Sustainable implementation success factors of AGVs in the brazilian industry supply chain management", *Procedia Manuf.,* vol. 39, pp. 1577-1586, 2019.
[http://dx.doi.org/10.1016/j.promfg.2020.01.284]

[101] M. Cummins, and P. Newman, "Appearance only SLAM at large scale with FAB-MAP 2.0", *Int. J. Robot. Res.,* vol. 30, no. 9, pp. 1100-1123, 2011.
[http://dx.doi.org/10.1177/0278364910385483]

[102] K. Konolige, and M. Agrawal, "FrameSLAM: From bundle adjustment to real-time visual mapping", *IEEE Trans. Robot.,* vol. 24, no. 5, pp. 1066-1077, 2008.
[http://dx.doi.org/10.1109/TRO.2008.2004832]

# CHAPTER 5

# Internet of Green Things (IoGT) for Carbon-Free Economy

**Sadiq Mohammed Sanusi[1,*], Singh Invinder Paul[2]** and **Ahmad Muhammad Makarfi[3]**

[1] *Department of Agricultural Economics & Extension, Federal University Dutse, P.M.B. 7156, Dutse, Nigeria*

[2] *Department of Agricultural Economics, SKRAU, Bikaner, India*

[3] *Department of Agricultural Economics and Extension, BUK, Kano, Nigeria*

**Abstract:** War, pollution, and a plethora of other threats are prevailing in the world on a daily basis. The globe, which has a population of more than 7.3 billion, is constantly harmed by human activity. Climate change is one of the world's most lethal problems because of these severe deformations. UN reported that 10000 people have died in extreme weather events like fire and floods in the past two years. Globally, $280 billion is lost to climate catastrophes from 2021 to 2022. While a single action won't be able to stop or slow down climate change, many tiny contributions from several professions will help it have an emotional effect. Scientists from all over the world are looking for ways to manage the transformation of the landscape in order to anticipate the dangers of climate change and, if possible, to reduce their effects on the future of the earth. Particularly with the Internet of Things (IoT), it is possible to slow the increase in global temperatures and cut back on hothouse emigration. The Internet of Things (IoT) encompasses more than just cutting-edge gadgets and intelligent machinery since it affects the state of the planet, from its climate to its financial resources. If we use technology effectively, it may be the instrument that saves the world. Since it involves everything from monitoring ozone levels in a meat packing facility to keeping an eye on public trees for banks, environmental monitoring is a broad activity for the Internet of Things (IoT). These findings mark the beginning of the process of developing several interconnected architectures that will support cutting-edge services and have greater effectiveness and flexibility.

**Keywords:** Carbon-free, economy, environment, GHG, IoGT.

* **Corresponding author Sadiq Mohammed Sanusi:** Department of Agricultural Economics & Extension, Federal University Dutse, P.M.B. 7156, Dutse, Nigeria; Tel: +2347037690123; E-mail: sadiqsanusi30@gmail.com

S. Gowrishankar, Hamidah Ibrahim, A. Veena, K. P. Asha Rani & A. H. Srinivasa (Eds.)

## INTRODUCTION

The United Nations (UN) introduced sustainable development goals (SDGs) after the 2015 millennium development goals (MDG) to address ecological, climatic, and health issues sustainably in order to support the future ecosystem for improved health in a sustainable community. In order to ensure the development of society, it is necessary to investigate the consequences of climate change using systematic methods and to develop indicators, frameworks, and tools to monitor and analyze various risk factors on health. For sustainable health, IoT must be combined with tools for detecting, communicating, monitoring, and decision assistance. According to Thilakarathne *et al.* [1], a variety of current and potential future climate change-related factors, such as rapidly varying ecological emissions, ozone exposure, temperature, air quality, and other weather-related events, pose a significant and varied threat to human health. The rising temperatures brought on by altered plant life, including the introduction of allergens, and rising carbon dioxide levels. As emissions increased, temperatures and sea levels rose, precipitation patterns changed, and a more extreme climate resulted. Disease transmission was sparked by contaminated surface water, food supplies, and deteriorated air quality, creating new concerns for human health [2]. Fig. (**1**) depicts the various health implications of a changing environment, which exacerbates current conditions and generates new demands.

**Fig. (1).** Green economy.

According to Rayan *et al.* [2], the impact of extreme weather on human health varies both momentarily and regionally and mostly affects the young, old, impoverished, and unwell. Indoor air quality is impacted by fungus and mold, heat waves, extremely high temperatures, and precipitation. While indoor moisture increases the prevalence of upper respiratory symptoms and asthma, high temperatures increase air pollutants and allergens, leading to rising hospital admission rates, ER visits, and moralities of children with asthma [3]. Similar to major floods, risky algal growth, and waterborne infections, significant precipitation leads to these conditions [4]. Due to breathing smoke, the increasing wildfires have a negative impact on the quality of the air outside and cause respiratory issues. Ozone levels also have a negative impact on the quality of the air outside [5]. In addition to more waterborne and food-borne illnesses, climate change has resulted in an increase in various disease vectors, including ticks that carry Lyme disease. Climate change is linked to poor mental health and increased stress, particularly after crises and displaced people [6].

Scientists from all over the world are searching for solutions to regulate environmental transformation in order to anticipate the hazards of climate change and, if possible, to reduce their effects on the planet's future. The transition of global industry to "Industry 4.0" status and the integration of information technology into every facet of human endeavor represent the most recent trends in the sector's growth [2]. The idea of the Internet of Things is one of them (IoT). With the help of "smart gadgets," which are connected to a worldwide network, you can track and assess the overall health of the ecosystem as well as find solutions to particular issues to reduce the detrimental effects of human activity. Most of the global SDGs-17 in total that were accepted at the UN Summit in 2015 have to do with environmental concerns: fighting climate change is goal 13 while goals 6, 7, 9, 11, 12, 14, and 15 are indirectly related. Two years later, the National Intelligence Council (NIC), the focal point of the American Intelligence Community's (IC) medium- and long-term strategic thinking, identified seven global trends that, in order to be effective, must be addressed by the year 2035 [2]. The environment, health, and climate change are all addressed in point seven. According to its norms, society will soon face issues that will alter living standards and living conditions, including more extreme weather, water and soil issues, glaciers melting, environmental pollution, and food insecurity. Global predictions indicate that tensions over climate change will only grow, necessitating swift and intelligent response [2, 7] (Fig. 1).

## MAJOR IMPACTS OF CLIMATE CHANGE ON HEALTH

The first official recognition of climate change as a serious hazard to humanity came with the signing of the Montreal Protocol in 1978. The Protocol was created as a result of observations made from space, opening the door for global environmental cooperation. Numbers speak for climate change more than three decades later, calling for a stance to steer the globe away from its repercussions. Since 1990, there have been significant increases in gas emissions, global temperatures have risen by 1.0°C since 2017, and sea levels have risen by around 20 cm since 1880 [8]. Humanity has watched Siberia and Australia engulfed in flames during the final months of 2019. The economy has been hampered by locally unusual weather conditions. Since the start of the novel coronavirus disease (COVID-19) pandemic, researchers have discovered a strong connection between the environment and human health, demonstrating the benefits of environmental deregulation. According to the World Health Organization (WHO), extreme temperatures, natural catastrophes, unpredictable rainfall patterns, and infectious diseases are all the ways in which climate change affects human health. Particularly for older adults, prolonged exposure to extreme heat has a significant negative impact on respiratory and cardiovascular health [9]. The increasing concentration of pollen and allergens in the air may make chronic illnesses like asthma or chronic obstructive pulmonary disorders (COPD) worse. Dehydration and electrolytic problems are more common in older people with heart failure or those using diuretics. Given this context, it is hardly unexpected that Europe's 2003 summer heat wave has been linked to at least 70,000 additional deaths [2]. There are immediate and long-term effects of natural disasters. Since the 1960s, their number has been rising on a global scale. Homes, vital infrastructure, and healthcare institutions are all destroyed when a hurricane or tsunami strikes a region. Large populations are compelled to move, where they must endure extended periods of substandard sanitation and scarcity of resources. Over 60,000 people every year die as a result of natural disasters. In underdeveloped nations, the associated mortality and morbidity are higher [10, 11].

Deregulation of the rainfall pattern might also have dire health consequences. Rainfall is essential to a community's ability to meet its sanitary and hydration demands. Rainfall also has a significant impact on the primary financial sector, especially on farming and agriculture. According to estimates, 500,000 children die each year from diarrheal diseases brought on by a shortage of clean water [2, 11]. In addition to depleting water supplies, floods can destroy homes and establish breeding grounds for disease-carrying insects, which raise infection risks. According to recent data, rainfall patterns may continue to be disrupted through the end of the twenty-first century, leading to droughts and famine. About 3.1 million people die each year as a result of hunger or undernutrition at the moment, and this number is predicted to rise [2]. Viruses, bacteria, protozoa, toxic algae, cyanobacterial toxins, and man-made chemicals are the principal causes of waterborne infections, which can be contracted through ingesting or inhalation of polluted water, eating tainted seafood, or engaging in recreational activities in polluted water [12, 13]. Due to a variety of environmental factors, such as storms, rain, floods, and tornadoes, as well as individual adaptability, the danger of waterborne infections increases and spreads. Having access to clean drinking water is a major challenge for billions of people throughout the world. The majority of gastrointestinal ailments in cities are caused by highly contaminated surface water and wells by numerous germs. This poses a major threat to public health. Increased global temperatures encourage the formation of toxic algae blooms that are dangerous to human health.

Because of COVID-19, worries regarding illnesses linked to climate change have increased. Numerous infectious diseases have connections to climate change, even though there are still many unanswered problems in this area. Water and sanitation are actually the subjects of the UN's sixth goal for sustainability. Recent climatological changes have led to an increase in the prevalence of snail-borne schistosomiasis in China. The development of insects like the Anopheles and Aedes mosquitoes, which serve as carriers of dengue fever and malaria, respectively, appears to be favored by climate change at the same time. Africa presently loses more than 400,000 people to malaria each year, and there are concerns that this number could rise [10]. Electrical storms are brought on by the rising humidity and may cause respiratory and cardiovascular diseases. Problems with psychological health, such as post-traumatic disorders, are exacerbated by natural crises. Longer hurricanes, tropical storms, and other severe weather events can cause stress that interferes with daily life and has an impact on one's health and well-being. Due to its ability to damage plastic, wood, and other materials as well as the environment, ultraviolet (UV) radiation poses a threat to both public and environmental health. Exposure to UV radiation can result in immune system deterioration, sunburns, skin cancers, cataracts, and ocular injuries [14]. Evidence estimates that the death toll of climate change may account for 250,000 more fatalities per year between 2030 and 2050, with 38,000 of those deaths coming from heat exposure in senior people, 48,000 from diarrhea, 60,000 from malaria, and 95,000 from under-nutrition in children [9]. While all people are at risk from climate change, the WHO says that children and communities living in remote locations, notably islands, are more susceptible [9]. It takes thorough study, effective policymaking, smart alliances, significant financial resources, and widespread awareness to address the effects of climate change on health.

## THE LOW CARBON ECONOMY AND ICT

People all over the world are facing a significant dilemma thanks to climate change. The Internet of Things (IoT) does, however, provide a variety of tools and resources that assist governments and enterprises in minimizing the negative effects of human activities on the planet [15]. Debates are also stoked in large part by climate change because it makes people consider scarcity. If people acknowledge the issue, they might have to forgo their comforts and pastimes in order to lessen their carbon footprints. This isn't always the case, though. The fundamental factor driving global warming and extreme weather is humankind's release of massive amounts of carbon dioxide into the atmosphere. Burning fuels like coal, natural gas, or petroleum results in the production of $CO_2$. Thus, even as we use electricity in more ways, we must simultaneously become drastically more energy-efficient and eliminate carbon from the electricity generating system (like powering vehicles). Electricity must be generated in the future using carbon-free, renewable resources like solar and wind. While still consuming less energy, IoT can assist in improving the speed, power, and efficiency of consumer goods. The decarbonization of our energy system, universal access to modern energy systems, infrastructure management, and the ability to respond to and solve climate change are all made possible by the Internet of Things.

IoT developments have made it possible for businesses, governments, and consumers to combat climate change since 2016 without compromising their convenience. More customers are expected to engage in activities that enable corporations and governments to gather Big Data and use it for environmental analysis. The future of environmental health and climate change is made possible by connected gadgets. 1.6 billion linked devices will be used by smart cities worldwide, according to Abd El-Mawla *et al.* [16] report from 2016. Additionally, according to a recent study by Ericsson (Fig. **2**), information and communication technology (ICT) might reduce greenhouse gas emissions by up to 63.5 Gtonnes by 2030. We may also witness, for the first time, a simultaneous decrease in the environmental footprint of companies as a result of energy savings and better solutions, given the increasing number of synergies that the IoT creates between various industries.

There is a necessity for increased IoT solution adoption if all those global leaders and industries continue expanding and be competitive [17]. Because of this, the Internet of Things may possibly be seen as a potent weapon in the battle against climate change, especially at a time when world leaders call for more feasible and scalable economic solutions to safeguard the environment. The globe will eventually be made up of many interconnected systems, allowing us to rely far more on intermittent sources of energy like solar and wind. Due to their intermittent nature, many experts believed that such sustainable sources could only make up a maximum of 10% of the world's total energy production. However, as the potential of the IoT has become more apparent, it now appears that up to 80% of the world's total energy production may eventually come from renewable sources. For instance, NRG, a massive commercial utility, works hard to meet its objective of reducing its overall carbon emissions by 50% by 2030, even though the company anticipates continuing to grow, and by 2050, it makes a pledge to cut $CO_2$ by an astonishing 90%, using emissions from 2014 as a baseline.

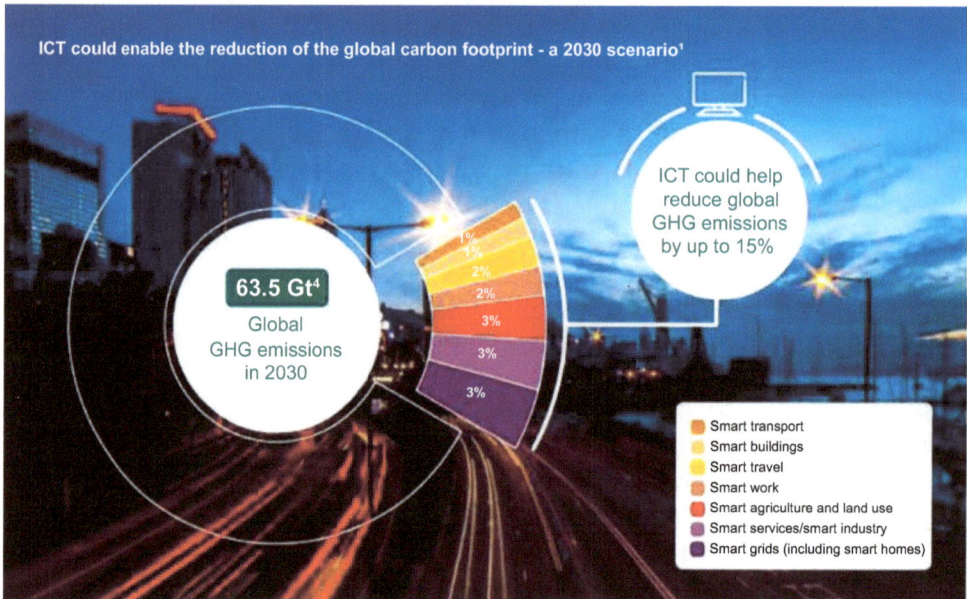

**Fig. (2).** ICT reducing global GHG by 2030 [Source: 16].

The process of combating climate change is perceived by world leaders as an overly complex challenge, but estimations and evaluations from communications equipment major Ericsson and environmental group Carbon War Room claim that advances in machine-to-machine communications, or IoT, can get us a significant amount of the way there, decreasing to as much as an 18% by 2030. The Internet of Things (IoT) opens the door to the development of an intelligent electric system that allows for significant supply and demand flexibility [18]. A responsive energy network that allows for real-time correlation of production and consumption data, thanks to IoT smart systems is also necessary. As a result of improved methods for forecasting demand, usage, and energy storage, power won't need to be generated solely when it is needed. We could thus get to the reality demonstrating that all of these enormous numbers of connected devices will provide various more approaches to improve human, environmental, and energy use efficiency and lower carbon emissions. IoT will revolutionize not just our lives but the entire world; it is so obvious [19]. One of the few sectors of the economy where new items tend to improve and consume less energy is technology.

## GLOBAL RISKS 2019 REPORT: FAILURE OF CLIMATE CHANGE MITIGATION AND ADAPTATION

The top 10 urgent threats facing the globe are presented in the Global Risks Report of 2019, which is in its 14th edition and was nearly created by 1,000 of the brightest decision-makers from the public and corporate sectors and civil society [16]. Nine out of ten respondents believe that this year will see an increase in political and economic tensions between major powers. The biggest concerns over a ten-year period are perceived to be climate change and weather policy failures. The "what-if" subjects covered in this year's study are incredibly in-depth and tackle numerous pressing issues, including weather manipulation, emotionally intelligent artificial intelligence, monetary populism, and other potential threats. The issue of emotions is addressed in the chapter on human causes and impacts of global hazards (The Global Risks 2019 Report as reported by [16]) to clearly call for larger and more impactful efforts in response to the rising levels of psychological stress worldwide.

Despite the differences between the two lists above, there are still some themes that tie them together. Three of the top five risks by likelihood and four of the top five risks by impact category were related to environmental issues, as seen in the table above. As seen in Table **1**, if we were paying close attention, we would also see that "failure of climate change mitigation and adaptation" comes in second on both the effect and likelihood lists, indicating that respondents are growing increasingly concerned about the failure of environmental policies. Risks related to the global economy, society and geopolitics are also causing a lot of worry in terms of possibility and impact. A surprising finding is that cyber-attacks appear in both top 10 lists, at number five for likelihood and seven for impact, while data fraud appears at number four for likelihood only. This demonstrates how technology is also involved in threatening global security, as we will demonstrate at the conclusion of this paper.

**Table 1. Top 10 global dangers in 2019.**

| Top 10 Dangers Ranked by Likelihood | Top 10 Dangers Based on Impact |
|---|---|
| Unusual weather conditions | Weapons of mass destruction |
| Failure to mitigate and adapt to climate change | Failure to mitigate and adapt to climate change |
| Natural catastrophes | Unusual weather conditions |
| Theft or fraud involving data | Water shortages |
| Cyber-attacks | Natural catastrophes |
| Environmental catastrophes caused by humans | Ecosystem destruction and loss of biodiversity |
| Massive uncontrolled migration | Cyber-attacks |

(Table 1) cont.....

| Top 10 Dangers Ranked by Likelihood | Top 10 Dangers Based on Impact |
|---|---|
| Ecosystem destruction and loss of biodiversity | Failure of the critical information infrastructure |
| Water shortages | Environmental catastrophes caused by humans |
| Bubbles in assets in a major economy | Infectious illness transmission |

Source: World Economic Global Danger Perception Survey, 2018-2019 [culled from 16].

Massive data breaches that exposed the personal information of millions or perhaps billions of individuals occurred last year, and cyber-attacks on both public and private organizations and enterprises persisted. In 2019, the majority of respondents (82%) anticipated that there would be a greater risk of cyber-attacks resulting in the theft of money and data, and 80% thought they would impede business operations. The societal risk was identified and given top priority on both lists under "a considerable drop in the available quality and quantity of fresh water, resulting in adverse consequences on human health and/or economic activity" water crises. It is positioned ninth for likelihood and fourth for impact. The paper forewarns the macroeconomic dangers the world faces as 2019 approaches. The dangers of "economic clashes between major powers" (91%) and "erosion of multilateral trading rules and agreements" (88%) are also expected to rise this year, according to the respondents. The authors emphasize the importance of financial market volatility and the pace of global expansion through 2018; this is supported by the most current international fund predictions. As a result, there is a 10% chance of a quality bubble developing in a very developed economy. These dangers are not independent or distinct from one another. As the following chart from the report examines, they are completely interrelated and each has the capacity to influence the others.

## INTERNET OF HEALTH THINGS (IOHT)

A strong health IoT architecture for addressing climate changes, including better air quality, more green urbanization, and fewer floods, could be supported by the application of IoT to healthcare in a way that sustainably develops both health and the environment in tandem with economic and societal growth [20, 21]. By tracking health outcomes across the full spectrum of a changing climate, providing data on health and environment that describe such outcomes, and facilitating the prediction, determination, and proper management of risk factors, health IoT holds the promise of enhancing sustainable health and well-being. Modern communication technologies like unique mobile health, innovative biosensors, remote and speedy diagnostics, disease-predicting models, telehealth, daily self-monitoring, and tracking approaches for better managing diseases and cutting-edge therapies are examples of IoT uses in the health sector [22, 23].

One of the most promising and forward-thinking areas for the achievement of environmental goals is the IoT idea. IoT technologies are currently available that enable analysis of the ecological state of many regions of our world [24]. Many of them have already been modified to work with management techniques for reducing nature's unfavourable effects in areas with high human density, particularly big and medium-sized cities. Special "smart" sensors that have been developed and put into use continuously gather data from which the appropriate judgments are made and actions are done to stop hazards brought on by some anomalous natural events and their harmful human impact [25]. Monitoring of the aquatic environment, air quality, seismic activity, illegal deforestation, peat land research, studies of the Great Barrier Reef, waste management, and even the wise use of public lighting are now commonplace.

Different sorts of tasks are involved in environmental protection, and in the majority of cases, the concept of IoT can be used to solve them in the proper way. There are still numerous intricate and understudied environmental issues, nevertheless. It will take some time to research, understand, and develop a neutralizing technique for their solution. Consequently, gathering the required data is the initial step in this direction. Because of this, there are now millions of "smart" devices connected to a single network that track the effects of human waste, mining products, and other ecosystems such as forests, rivers, seas, and ecosystems. Scientists create and enhance mobile applications for reading and analyzing data as well as specialized environmental sensors for collecting in line with the specifications. The scope of their skills is fairly broad, ranging from registering radiation levels to evaluating environmental characteristics (such as air quality, temperature, humidity, and carbon dioxide content) and assessing the amount of nitrates in food. The degree of data accuracy is increased by using RFID, Wi-Fi, Bluetooth, GPS, Zigbee, LoRaWAN, and NB-IoT modules. Personal sensors allow for several information reception methods, and they communicate information to both personal computers and smartphones for additional processing [2].

## IOT APPLICATION IN COMBATING CLIMATE CHANGE

Advances in machine-to-machine connections, or the Internet of Things (IoT), are one approach to accomplish this mission of battling climate change. The IoT can significantly contribute to as much as an 18% reduction by 2030 [26]. IoT is a network of things that can connect to the internet, process data, and exchange it with each other to enable smart solutions [27] (Van *et al.*, 2022). These objects are equipped with a variety of sensors, software and applications, network connectivity, and computing capabilities. The Internet of Things (IoT) is already providing uncommon prospects for solving a number of environmental problems, such as clean water, landfill trash, deforestation, and air pollution, and will ultimately help lessen the environmental effects of human activities [27]. According to a survey, the Internet of Things (IoT) industry will grow significantly around the world since all major corporations recognize the critical role that IoT plays in protecting both the environment and human lives [28]. The following points will discuss many life domains where IoT offers practical solutions for it to save the environment and help solve the problem of climate change in order to provide a better representation of the usefulness of IoT in combating climate change.

## Agriculture

Due to the fact that the carbon emissions from agriculture are among the highest on the globe, the industry has worked for many years to lessen its own impact on climate change. Since the agricultural industry is one of the most crucial, it is looking to the Internet of Things for solutions that will enable smart farming and precision agriculture, among other initiatives, to considerably increase its sustainability. Precision agriculture will advance through the use of sensors to gather, process, and transmit data during agricultural operations. IoT gadgets will also improve the accuracy of weather forecasting methods, allowing farmers to use their resources more effectively and produce less waste. As a result, the living standards of low-income and emerging nations will be improved, and the environmental impact of agricultural activities will be diminished. The International Telecommunication Union (ITU) discovered that the number of homes in isolated rural areas with access to electricity and the internet is steadily rising, which strengthens the possibilities for incorporating technology into agriculture in many ways that reduce the consumption of insecticides, fertilizer, and water to the absolute minimum, as well as weather forecasting systems. The general elements of a greenhouse environmental monitoring system based on WSNs and IoT are described in Fig. (**3**).

## Stopping Illegal Logging and Deforestation

The detection and reporting of illegal logging and forestry operations that contribute to climate change are now frequently done with the help of IoT-enabled devices [29] (Ebiesuwa *et al.*, 2022). One of the main contributors to climate change, according to theory, is deforestation. One billion trees must be planted a year to recover, according to a study, and this is only feasible with the aid of an automated system. In order to effectively monitor deforestation, IoT and drones are installed, built, and used. This lessens the need for labour, saves money, and facilitates the use of contemporary agricultural techniques for planting and upkeep. For example, the San Francisco-based start-up rainforest connection project uses discarded cell phones to create solar-powered listening devices that are affixed to trees to monitor and identify remote illicit logging activity. As a result, areas like the Amazon forest and the Indonesian region of Kalimantan that are prone to deforestation would receive better protection.

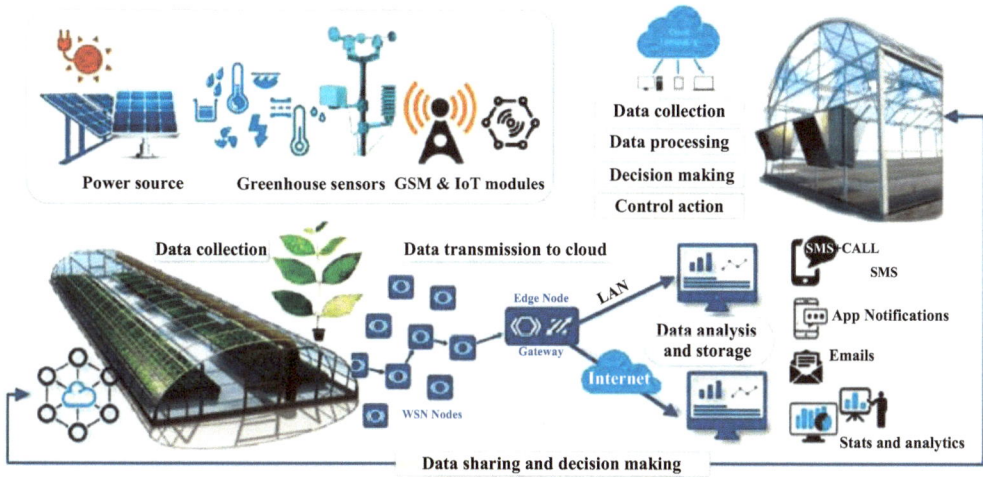

**Fig. (3).** Environmental monitoring of greenhouse based on IoT and sensor wireless networks [Source: 16, 26].

## Smart Cities

According to the most recent adjustment of the UN population projections, 70% of our species is expected to live in urban areas by 2050, which will transform how humans interact with their surroundings. The next stage in integrating structures, residences, workplaces, warehouses, and various forms of public infrastructure is to create smart buildings and cities [30]. In order to minimize energy waste and the building's carbon footprint, operations, including lighting, power, waste collection, and alarm systems, as well as many other technologies, can be connected to centralized management systems. One day, a smart city could guide you through its streets on a route that best avoids stopping at traffic lights or in places with a lot of traffic using IoT-powered intelligent transportation systems. In order to truly make the smart city intelligent, these systems would also be driven by the data that they and other linked systems generate and exchange. This data would then probably be used by a central management system. The aforementioned IoT applications could all be used to assist in counteracting the effects of climate change and keep any extreme weather or conditions to a minimum as they become more significant for everyone on the planet (Fig. **4**).

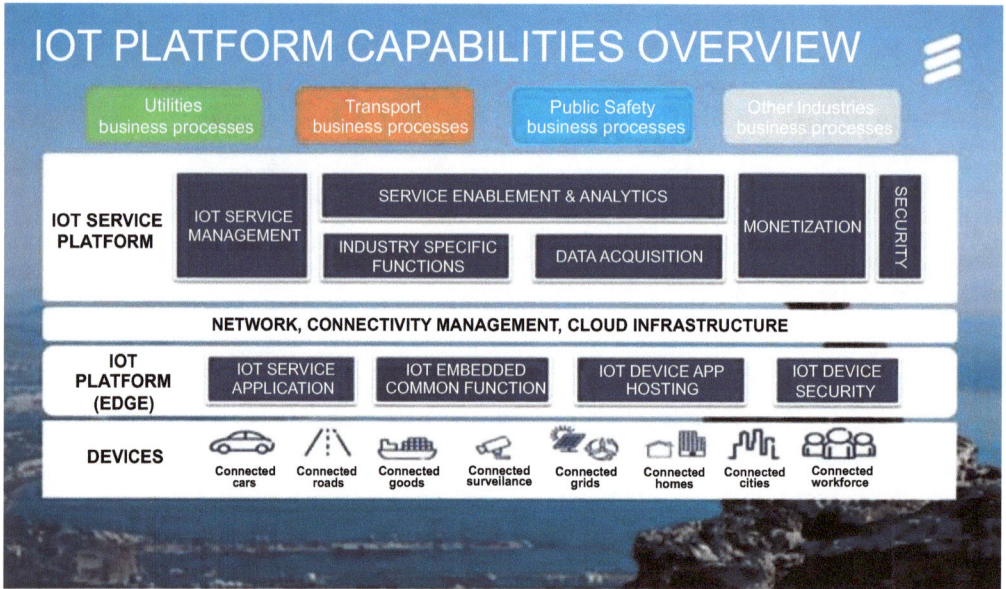

**Fig. (4).** Ericsson smart city vision [Source: 16].

## Utilities

The Internet of Things (IoT) for utilities plays a significant part in assisting various utilities by enabling them to employ a quick, responsive energy network that uses predictive analytics to match and accomplish energy generation with demand, which results in storing excess energy rather than waste it. Smart metres may also gather data on each building's energy consumption and relay that information back to utilities to assist in load balancing, which will reduce waste as well. All of this can minimize the quantity of fossil fuels utilities must utilize to produce energy, which lowers their carbon footprint.

## Waste Control

Early IoT technology projections suggested that e-waste would rise, but the answer included addressing this issue. IoT now offers a fantastic approach to control garbage and is very important to the recycling process [8]. When it is time to repair or recycle, the attached sensors will be able to let the consumer know. In order to prevent premature trashing, they also let users know which parts can be reused. IoT solutions can be created to automate timely disposal systems for other waste management.

## Transportation and Traffic

By assisting drivers in finding parking spots more quickly and avoiding congested roads to cut down on the amount of time they spend driving, IoT-powered apps help the world minimize carbon dioxide emissions from cars. Additionally, it allowed vehicles to plan their trips around times when the roads were clear, which helped to use less fuel. For example, a project in Los Angeles employs IoT technology to coordinate traffic lights to enable more efficient traffic flow, lowering greenhouse gas emissions and saving more than 35 million gallons of gasoline annually. Additionally, in order to improve garbage collection and lower carbon emissions, IoT-enabled trash cans have been installed to alert collectors when they are full. Another benefit of IoT data collection was that it allowed governments and transportation corporations to cut out unnecessary routes, which reduced operational costs and carbon emissions while also saving money. Additionally, data can be used to create better routes, bringing in more passengers for public transportation companies while also providing a more environmentally friendly alternative to people's less efficient personal vehicles, as stop-and-go traffic increases energy usage, considering the Advantech Intelligent Highways Project [16], which is described in detail in Fig. (**5**).

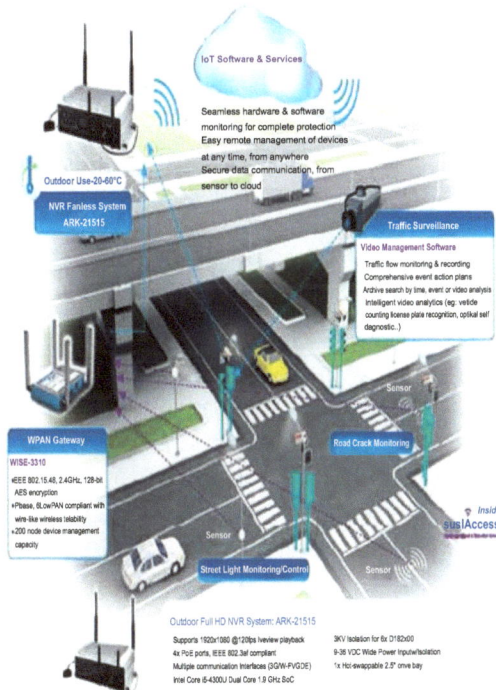

**Fig. (5).** Intelligent Highways of ADVANTECH Project (**Source:**www.advantech.net.au).

## Data About the Climate and the Environment

The process of assessing the impact of global climate change depends on climate models (Fig. **6**). However, even the most advanced models still have limitations, and more data is required to effectively simulate the climate and predict changes. The Internet of Things paradigm is excellent at generating information, therefore, gathering climate data will be made possible by it [31]. Around the world, there are already sensors for temperature, humidity, and precipitation. However, IoT devices can provide more reliable data from many more sources. These gadgets can more precisely monitor ocean temperatures and sea levels than ever before, and the information will be very useful for reducing possible changes brought on by climate change. Scientists from all around the world will be able to notice minute changes and modify their models as a result of the precision provided by IoT devices. Due to the fact that most individuals are passionate about protecting the environment in some way, numerous programmes have been established to reduce emissions, pollution, illegal logging, and other harmful or polluting activities. These causes have utilized IoT devices to great effect as they have become more widely available. Using sensors connected to trees that listen out for and report unlawful logging, one example illustrates efforts to minimize deforestation, which is responsible for 15% of all worldwide carbon emissions. Another illustration has a similar setup, but it is being used to spot unlawful poaching. The ability to negotiate more safeguards for these places to help fight climate change and deforestation is thought to be made possible by these kinds of technology, which allow groups to view this data in real-time. Observing how IoT combats pollution is another extremely intriguing example.

## Automated Buildings and Energy Storage

One of the key reasons promoting climate change is thought to be society's energy consumption [32]. This is because power plants produce a lot of pollution, like $CO_2$ emissions, or people use central heating. It is important to realize that a significant portion of energy consumption is not just a result of human demands, but also of ineffective management and lack of foresight. Due to inadequate data analysis, we really consume more energy than we actually need. Keeping track of people's energy use enables waste reduction. Consider the thermostats from Nest, Honeywell, and Lenox. While an unprogrammed thermostat can waste 20% of the energy used for heating and cooling, the problem is addressed by a smart thermostat, which learns your habits and makes energy-saving adjustments on its own. Many examples of smart thermostats are shown in Fig. (**7**). With something as basic as using an app to switch off the lights or with apps like IFTTT, which connect to a variety of systems, we can see how IoT saves energy and reduces carbon footprints. An intelligent energy storage system can be used to programme, monitor, and regulate IoT-enabled devices so that they only use energy when necessary.

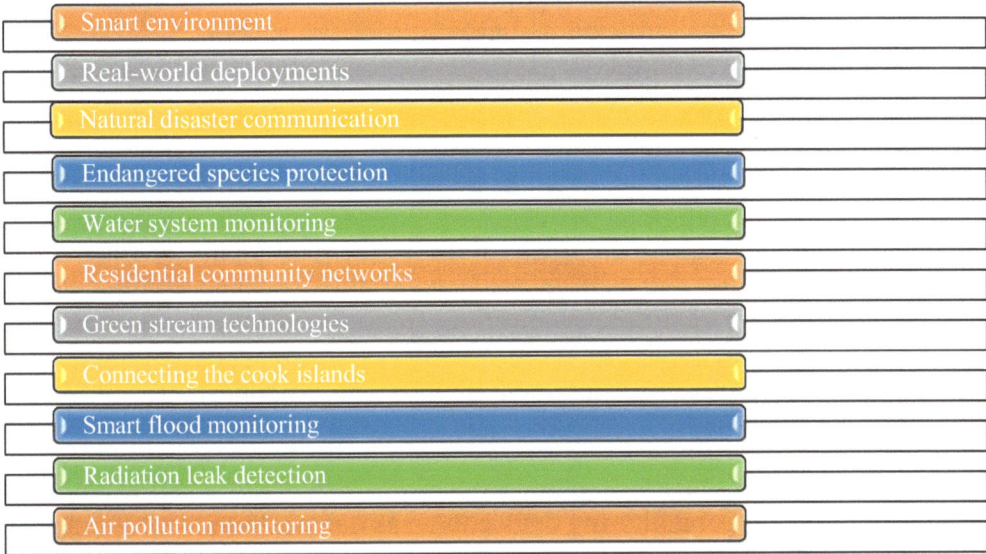

> Smart environment
> Real-world deployments
> Natural disaster communication
> Endangered species protection
> Water system monitoring
> Residential community networks
> Green stream technologies
> Connecting the cook islands
> Smart flood monitoring
> Radiation leak detection
> Air pollution monitoring

**Fig. (6).** IoT environmental applications.

**Fig. (7).** Smart Thermostats (Source: bestsmartthermostatreviews.com).

## IoT-Based Environment Solutions Examples

The IoT sensing technologies show promise in timely monitoring water quality and flow, alerting systems, and other treatment methods because they could enhance water quality and mitigate or prevent illnesses related to water pollution, which would inform choosing a suitable purifying approach in accordance with the identified contagious microorganisms. For example, India provides clean water *via* curing water in rural regions using reverse osmosis (RO) technology, sensor networks, and IoT-integrated intelligent metres [2]. Bangladesh employs arsenic biosensor networks to monitor water quality, whereas China uses IoT-integrated sensors installed in various locations of the water supply system to track water flow [33]. Kenya employs an intelligent hand pump for water instead of the faulty conventional ones because it uses sophisticated accelerometers with built-in 3G radios for real-time monitoring, limiting disruptions, and providing stable water delivery service [34]. To monitor and enhance personal hygiene practises like washing hands after using the restroom, Indonesia uses water flow and movement sensors [35].

The Air Quality Egg sensor, designed to measure air quality, is an example of a "smart" environmental monitoring device that has demonstrated value and gained popularity among consumers [2]. You may determine the amount of air pollution directly in the user's home or business and throughout the city using the data gathered by all devices linked to the network and displayed in real time on a dedicated website. America and Europe both use this technology, and poorer nations are starting to catch up. Additionally, successful devices like Speck, Sensordrone, and iGeigie have emerged [36]. Household garbage is the true disaster of modern cities. Bigbelly, a "smart" container, was created a few years ago to address this issue. The fifth generation of seals is aiming to promote cleanliness throughout the world as part of its ongoing technical progress. A clever solar-powered waste compactor is the HC5 Bigbelly. It has a sensor that keeps track of its fill level and wirelessly transmits the information to the appropriate city agencies. Streamlining waste management processes may boost productivity, maintain public spaces clean and green, and increase power availability. Special teams can arrange waste pickup and swiftly clean containers using the data collected from Bigbelly [37].

With the help of the IoT, numerous nations are attempting to improve the reliability and environmental friendliness of their cities' infrastructure. Studies in science claim that roughly 6% of the carbon dioxide in the atmosphere comes from street lighting. Consequently, some nations are working to increase the efficiency of their electric lighting systems in an effort to lower their emissions. For instance, Copenhagen's streets now have "smart" lamps, thanks to Denmark.

They use sensors to keep an eye on whether automobiles or people are using a particular stretch of the road, to evaluate the weather, and to modify the lighting's brightness and, consequently, the amount of carbon dioxide emissions based on this information. The new lighting network also presents a number of potential for future service connections, including those for air quality, noise, and video surveillance cameras, which can further enhance public safety and quality of life while transforming Copenhagen into a true "Smart City" [2]. To enhance street lighting, the American business TCS Digital Software & Solutions Group (Tata Consultancy Services) has released ground-breaking cloud software called Intelligent Urban Exchange (IUX). Both common streetlights and indoor LED lamps are compatible with this approach. Smart lights can react to rapid movements made by people by raising brightness to warn of danger and decreasing it in the case of low pedestrian activity, thanks to specialized sensors. In accordance with the climate and the amount of smog/pollution in the air, smart lighting can also be automatically adjusted. The problem of energy optimization should, therefore, be successfully addressed by the new intelligent street lighting software. To stop illicit logging from occurring in the Brazilian Amazon, a specific initiative has been designed. A unique mobile device called Invisible Track can identify individual trees in the Brazilian Amazon protected region in a certain order. When this gadget is within 20 miles of a cellular base station, it has a communication module that alerts the Cargo Track Operation Center to its location. One of the modified trees is reportedly being delivered to the Brazilian Environmental Protection Agency. In order to catch and detain thieves, whether they are in a sawmill or a rainforest, law enforcement officials receive real-time location information. This gadget can function independently for more than a year without recharging and is ideally suited to the Amazonian climate. It operates in places with low signal levels and little volume thanks to radio communication technology (RED), which makes it almost invisible. Even in the most remote locations, RED technology and Cargo Track geo-location algorithms deliver great location accuracy. This technology has demonstrated its efficacy right from the start of its use.

Thales, a leader in digital security, was honoured by the Internet development group (IDG) as the winner of the "Computer World" in the category "World Good" for its dedication to promoting and developing cutting-edge IoT solutions that solve such global concerns. The successful experience makes it more likely that one of the most prized and significant resources on earth, the planet's forests, which serve as its lungs, will be preserved [2]. The Great Barrier Reef, which runs the length of the Australian continent and was named a UNESCO World Heritage Site, is a habitat for a vast array of living things and has a big effect on the local ecology. The IrrigWeb-automatic farm irrigation system was created by Dr. Eric Wang to achieve optimal preservation, and it was later enhanced by a WiSA system. As part of a pilot study on a sugar cane farm in the Burdekin, researchers Wang *et al.* [38] successfully created and implemented this approach. The Great Barrier Reef's flora and fauna are negatively affected by run-off, therefore, this technology is intended to automate the irrigation system for fields and limit both the amount and possibility of ocean entry. Data are gathered on the Great Barrier Reef's biochemical condition using buoys that are fitted with specialized sensors. Australian conservation organizations use sensor data to assess the degree of harm to coral reefs, fish movement, and population health caused by various bacteria [38].

A new digital solution called "smart spraying" was created in 2017 as a result of the collaboration of two recognized global firms, Bayer and Bosch. With the support of such a collaborative research agreement, farms would be transformed into digital farms, plant protection goods could be used more effectively, and herbicides and insecticides could only be used where they were truly necessary. High-efficiency sensor technology, "smart" analytical tools, and a selective spray system are the main areas of concentration for the company's development [2]. What happens if the farmland is situated in areas far from cities and communication infrastructure? Conventional mobile technology can only partially address this issue because building traditional communication links is relatively expensive. Technology built on LoRaWAN comes to the rescue. Kovalchuk *et al.* [39] researched new potential applications and features of their use. In their research, they looked into the possibility of using the LoRaWAN protocol to connect moisture, temperature, pressure, direction, and wind speed sensors to Internet gateways without having to pay for cellular communications, additional power, or the installation of intricate Wi-Fi networks in the field. IoT holds forth the prospect of actively monitoring ozone and UV radiation levels on a wide scale, thereby creating frameworks for future decision-making. By measuring the amount of absorbed ionizing radiation, a sensor known as a dosimeter, for example, might determine UV exposure [40].

## IOT ENVIRONMENTAL TECHNOLOGY PROJECT CASE STUDY BY ERICSSON

(Fig. **8**)'s "Smart Grid," as described by Ericsson Research, is expected to reduce greenhouse gas emissions by 3.9 percent by 2030 [16]. In addition to offering significant environmental advantages, this evolving IoT-enabled energy delivery network can monitor and adapt to local variations in energy demand. The smart electricity metre, a minor part of this smart grid, allows for real-time two-way communication between utility and user, facilitating the latter's ability to satisfy energy demand with less waste by coordinating energy generation and conservation. Customers will be encouraged to use less energy during busy hours as a result. The World Green Building Council, for instance, is working hard to coordinate regional efforts to make sure that all new buildings operate at zero carbon in order to lessen the impact of carbon on daily economic activity. The Council's primary objectives, such as lowering energy consumption, producing renewable energy, and performing closed-loop measurements of carbon consumption and waste, are all supported by IoT sensors that are currently accessible. According to the Ericsson Project, the public and commercial sectors might both benefit from IoT-powered smart services that would reduce carbon emissions by 3% by improving efficiency and reducing their dependency on throwaway materials. Smart agriculture (which reduces the consumption of fertilizer, pesticide, and water) may now be implemented in modernizing nations much more quickly than was previously conceivable because of low-cost, low-power IoT devices.

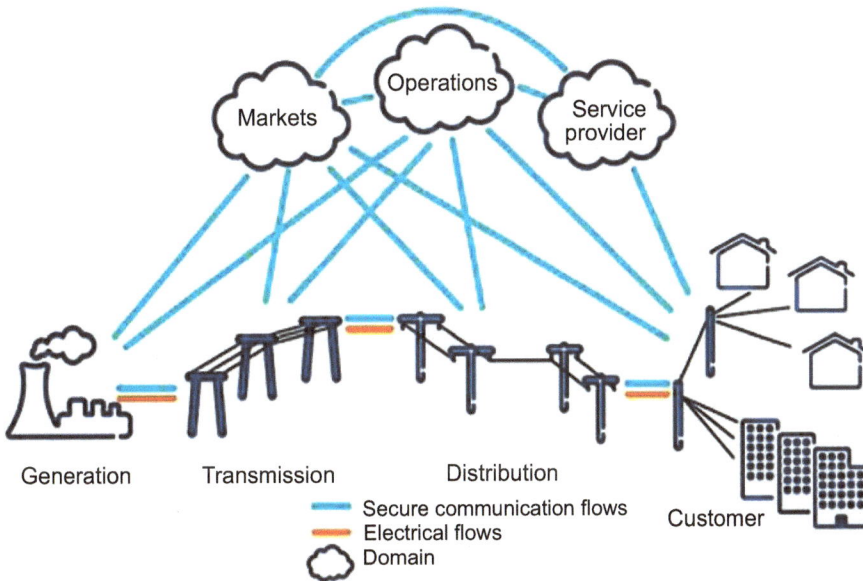

**Fig. (8).** Ericsson Smart Grid (Source: www.ericsson.com).

Additionally, smart agriculture could help with a further 3% reduction, according to Ericsson Research. Improved cow monitoring is one example of smart agriculture practices that can be used to spot sick animals in the herd based on their behaviour and position [25]. This information can then be used to screen, treat, and get rid of unhealthy animals. The world will benefit from increased production of meat and dairy products, decreased overuse of antibiotics, and chances to lower greenhouse gas emissions from unhealthy animals thanks to this method. The Ericsson Project demonstrates yet another method for cutting fuel use by up to 15%: implementing IoT monitoring and rerouting for marine cargo shipping. Early preventative maintenance can lessen the requirement for redundant fleets by allowing for on-the-spot repairs rather than five-week overhauls.

William Nordhaus and Paul Romer received the Nobel Prize for their research on the economics of global warming as a result of the UN's climate change report [16]. To implement a tax that reflects the social cost of carbon emissions by humanity, they advise world leaders and politicians. If not, in the coming decades, the world will experience severe political upheaval, sickness, flooding, and food and water shortages. Although the US now views this kind of drastic policy change as political poison, several other major economies across the world, including as China, Canada, and some regions of Europe, have already begun to introduce carbon pricing. IoT technologies will play a key role in helping people realize that there is a significant financial benefit to being able to offer electricity in locations where they can do so without paying the tax. Researchers from Ericsson concluded that by 2030, ICT may reduce greenhouse gas emissions across all industrial sectors by up to 63.5 gigatons, or 15%. The following are a few IoT solution examples provided by Ericsson:

a. *Electricity Distribution*: Smart metering, smart grid, and small-scale renewable energy sources in homes and businesses will assist in cutting losses and lower energy use.
b. *Services and Industry*: Innovative approaches to governance, healthcare, and education will lower the cost of the service sector and foster socioeconomic growth in underdeveloped regions of the world.
c. *Transportation*: By enhancing traffic flow, finding parking spots faster, and encouraging a shift toward more environmentally friendly options like public transportation, smart travel solutions will assist in achieving lower emissions ratios. The globe will witness a wide range of innovative new approaches to decrease waste and achieve reduced greenhouse gas emissions across every business by financially disincentivizing carbon usage. Without a doubt, the Internet of Things will be a powerful ally in the battle because of its efficiency. All of this entails avoiding countless years of excessive and inefficient fuel, water, and soil additive consumption.

## INDUSTRIAL INTERNET OF THINGS (IIOTS) IS THE FUTURE

Manufacturers, utilities, agricultural producers, and healthcare providers are implementing IIoT primarily to boost productivity and efficiency through smart and remote management, as well as to be environmentally friendly and live in harmony. As an illustration, Thames Water [41], the biggest provider of drinking and wastewater services in the UK, uses sensors, real-time data collecting, and analytics to predict equipment breakdowns and give quick reactions to urgent circumstances, including leaks or unfavourable weather occurrences. More than 100,000 smart metres have already been installed by the energy company in London, and it plans to provide smart metres to every customer by the year 2030. This technique has already saved London an estimated 930,000 litres of water daily after 4,200 leaks on customer lines were discovered. Another illustration is the Mitsubishi Chemical Plant in Kashima, Japan, where the installation of 800 HART devices for real-time process control has improved production performance while preventing a $3 million shutdown [16] and saving US$20 to $30,000. Through the deployment of a dense network of sensors, agriculture controlled by IIoT can assist farmers in measuring various agricultural variables, including soil nutrients, fertilizer used, seeds planted, soil water, and temperature of stored produce, improving productivity by almost doubling [25]. Agricultural IoT is being promoted by businesses like Microsoft (Farm- Beats project), Climate Corp, AT&T, and Monsanto [16]. The healthcare industry may also be greatly impacted by IIoT. False alarms, sluggish responses, and incomplete information are still key contributors to avoidable patient suffering and death in hospitals. Hospitals may considerably overcome these constraints by integrating distributed medical devices with IIoT technologies, enhancing patient safety and satisfaction while using resources more effectively. Enhancing productivity, security, and working environment for employees are further potential offered by IIoT. The use of unmanned aerial vehicles (UAVs) enables, among other things, the inspection of oil pipelines, the use of sensors to check food quality, and the reduction of worker exposure to noise and potentially harmful gases or chemicals in industrial settings. For instance, Schlumberger is already using autonomous marine vehicles to monitor subsea conditions. These vehicles can cross oceans and collect data for up to a year without needing fuel or a crew because they are propelled by wave energy. Mining companies may significantly reduce safety-related mishaps and increase productivity and profitability by using IoT-powered remote monitoring and sensing. For instance, Rio Tinto, a major mining firm, plans to use its automated operations in Australia to preview a future where all of its mines are more productive and there is less need for human miners. In addition to enhancing enterprises, IIoT is an environmentally benign technology that benefits both the environment and all of humanity.

## Challenges of Industrial Internet of Things (IIoT)

Since IIOT has a bright future, more money must be put into addressing its current major problems. It will be a step toward protecting the environment and the planet if these problems are resolved. The most recent IIoT challenges are listed below.

## Energy Savings

Environmental energy sources, such as thermal, solar, vibration and wireless radiofrequency (RF) energy sources, can really be used to generate energy. The availability of the corresponding energy source is necessary for harvesting from such environmental sources. The IIoT does not instantly benefit from the several energy-efficient wireless sensor networks (WSN) strategies that have been presented in recent years. Applications using the IIoT frequently require a dense deployment of multiple devices. Sent data can be in a queried or continuous form, both of which can use a lot of energy in a dense deployment. Thus, in the IIoT, green networking is essential for lowering power consumption and operating expenses. Making the most of surveillance and environmental preservation will reduce pollutants and emissions. Several energy-efficient design techniques are used by LPWAN IoT technologies to achieve low-power operation.

➤ A star topology, which eliminates the energy used for packet routing in multi-hop networks.

➤ Offload the complexity to the gateway to keep the node architecture simple.

➤ By using narrowband channels, you can increase transmission range while reducing noise.

➤ Communication standards that are lightweight.

➤ Using wireless transceivers with minimal power.

## Performance in Real-Time

For the support of applications that are mission and safety-critical, IIoT devices are frequently placed in loud settings, and they must adhere to strict timing and reliability criteria for the timely collection of environmental data and the appropriate delivery of control choices. Thus, the end-to-end (e2e) deadlines of the real-time sensing and control operations carried out in the system are frequently used to gauge the QoS provided by the IIoT. To achieve the appropriate QoS in the IIoT, time-slotted packet scheduling is essential. It is now significantly more challenging to achieve the needed real-time performance due to the increasing rise of IIoT applications, particularly in terms of their scale and complexity [42]. The challenge is made worse by the fact that the majority of IIoT must deal with unforeseen disruptions. Unexpected disruptions can be divided into two categories: external disturbances from the monitored and controlled environment (such as the identification of an emergency, abrupt pressure or temperature changes), and internal disturbances from the network architecture (*e.g.*, link failure due to multi-user interference or weather related changes in channel SNR).

Numerous centralized scheduling systems have been proposed in response to various internal problems. Several papers have been written about critical control systems that can adapt to outside disturbances. As a result, the majority of them are not very scalable. Distributed resource management is not a brand-new idea.

In reality, the wireless network community has looked into distributed techniques quite a bit. Real-time e2e restrictions, however, are often not a focus of these studies. It's still a challenge to guarantee finite response times for concurrent disturbances.

## *Interoperability and Coexistence*

There will be a lot of coexisting devices deployed in close proximity in the limited spectrum due to the rapid expansion of IIoT connectivity. The impending problem of coexisting in the congested ISM bands is brought up by this. In order to keep devices functional, interference between them must be managed. IIoT devices today and in the near future will probably have minimal memory and intelligence to prevent or reduce interference. Each IIoT technology has unique qualities that could provide additional difficulties. Future IIoT devices must be able to detect, identify, and eliminate external interference in order to ensure peaceful coexistence. Currently, some research on identifying interference on IIoT devices using spectrum sensing has been described, but much of the work falls short since a very long sample window is required and existing solutions demand far more RAM than what is provided in IIoT devices today. It is crucial to place more emphasis on researching and comprehending the complex radio environment where many of these IIoT devices will be deployed if research into error-correcting codes for IIoT devices is to be effective [43]. Multimode radios, software flexibility, and cross-technology communication are three dimensions in which the issues posed by device diversity in the IIoT can be handled.

## *Privacy and Security*

Another important IIoT risk is security. IIoT is a generally resource-constrained communication network that heavily relies on low-bandwidth channels for communication among lightweight devices, considering CPU, memory, and energy consumption. Traditional security measures, such secure protocols, lightweight cryptography, and privacy assurance, are therefore insufficient to defend the intricate IIoT networks. It is possible to study current encryption methods from industrial WSNs before implementing them to create safe IIoT protocols. For example, limited memory and processing resources preclude the employment of resource-demanding crypto-primitives, such as Public-Key Cryptography (PKC). Applications that include enormous data transmission and real-time demands provide a more serious difficulty. One may argue in favour of a comprehensive strategy [44] to address privacy and security issues in the IIoT. It follows that throughout the entire life cycle of the systems and goods, considerations such platform security, secure engineering, security management, identity management, and industrial rights management must be made. When developing secure IIoT infrastructure, there are a number of security properties to take into account [16]:

➢ IIoT devices should be tamper-resistant against physical attacks such passive secret stealing and unauthorized reprogramming while still allowing authorized users to upgrade the device's security firmware.

➢ Data should be kept encrypted to maintain confidentiality and safeguard the storage of IIoT devices from attackers.

➢ To maintain confidentiality and integrity, the IIoT device communication network needs to be secured.

➢ To ensure that only parties with permission can access the IIoT resource, the IIoT infrastructure needs effective identity and authorization methods.

➢ Despite malicious users physically damaging the equipment, the system should continue to function normally. This ensures the IIoT's resilience.

For IIoT devices, symmetric-key cryptography typically offers a simple solution. Nevertheless, symmetric-key encryption has significant drawbacks with regard to key management and storage, particularly for limited capacity devices. In addition, if one IIoT device is compromised, all other keys may be exposed. Generally speaking, public-key cryptography offers more secure features and requires less storage, but it has a large processing overhead because of the complicated encryption. Therefore, a significant challenge for IIoT security is to reduce the overhead of complicated security protocols for public-key cryptosystems.

## IOT FUTURE PERSPECTIVES FOR CARBON-FREE ECOLOGY

The advancement of IoT is intended to benefit the planet's ecological situation [45]. Although its potential is still untapped, IoT technology has the ability to give humanity fresh approaches to environmental issues. Scientific evidence shows that IoT devices socialize more quickly the more network connections they have [46]. In the IoT ecosystem, cloud technology is crucial. Increased computer power and storage space are possible with cloud computing [47]. Additionally, sensors can be utilized anywhere, and cloud computing services can process the data they collect. The data will then be readily accessible to many people thanks to cloud technologies. By means of a shared social network, IoT devices and customers will be linked (Social IoT-SIoT). An examination of the state of the environment can be provided by such environmental monitoring based on the SIoT platform's capabilities, which is more precise and efficient than current techniques. Users will be able to learn really important information about the earth's natural processes in real time as a result.

## CONCLUSION

Currently and possibly in the future, climate change poses a serious and diverse hazard to human health. In order to ensure the development of society, it is necessary to investigate the consequences of climate change using systematic methods and to develop indicators, frameworks, and tools to monitor and analyze various risk factors on health. Rigorous study, effective policymaking, innovative alliances, significant financial resources, and widespread awareness are needed to address the effects of climate change on health. One of the most promising and forward-thinking areas for the achievement of environmental goals is the IoT idea. IoT technologies are currently available that enable analysis of the ecological state of many regions of our world. For sustainable health, IoT must be integrated with tools for sensing, communicating, monitoring, and decision assistance. Therefore, there is a need to put more emphasis on developing and implementing sustainable solutions.

## REFERENCES

[1]     N.N. Thilakarathne, M.K. Kagita, and W.D. Priyashan, "Green internet of things: The next generation energy efficient internet of things", In: *Applied Information Processing Systems.* Springer: Singapore, 2022, pp. 391-402.
[http://dx.doi.org/10.1007/978-981-16-2008-9_38]

[2]     R.A. Rayan, I. Zafar, C. Tsagkaris, and I. Romash, "Internet of Things for mitigating climate change impacts on health", In: *Artificial Intelligence and Internet of Things.* CRC Press, 2021, pp. 317-330.
[http://dx.doi.org/10.1201/9781003097204-14]

[3]     Y.A. Qadri, A. Nauman, Y.B. Zikria, A.V. Vasilakos, and S.W. Kim, "The future of healthcare internet of things: A survey of emerging technologies", *IEEE Commun. Surv. Tutor.,* vol. 22, no. 2, pp. 1121-1167, 2020.
[http://dx.doi.org/10.1109/COMST.2020.2973314]

[4]     C.J. Gobler, "Climate change and harmful algal blooms: Insights and perspective", *Harmful Algae,* vol. 91, p. 101731, 2020.
[http://dx.doi.org/10.1016/j.hal.2019.101731] [PMID: 32057341]

[5]     F. Li, P. Newman, S. Pawson, and J. Perlwitz, "Effects of greenhouse gas increase and stratospheric

ozone depletion on stratospheric mean age of air in 1960–2010", *J. Geophys. Res. Atmos.,* vol. 123, no. 4, pp. 2098-2110, 2018.
[http://dx.doi.org/10.1002/2017JD027562]

[6]     K. Hayes, G. Blashki, J. Wiseman, S. Burke, and L. Reifels, "Climate change and mental health: risks, impacts and priority actions", *Int. J. Ment. Health Syst.,* vol. 12, no. 1, p. 28, 2018.
[http://dx.doi.org/10.1186/s13033-018-0210-6] [PMID: 29881451]

[7]     National Intelligence Council, *Global Trends: The Paradox of Progress,* 2017. Available From:https://espas. secure.europarl.europa.eu/orbis/document/global-trends-paradox-progress

[8]     G. Singh, G. Kumar, V. Bhatnagar, A. Srivastava, and K. Jyoti, "Pollution management through internet of things: A substantial solution for society", *Humanities & Social Sciences Reviews,* vol. 7, no. 5, pp. 1231-1237, 2019.
[http://dx.doi.org/10.18510/hssr.2019.75162]

[9]     WHO, *Climate Change and Health.,* 2018. Available From:https://www.who.int/news-room/fac--sheets/ detail/climate-change-and-health

[10]    T.C.M. de Sousa, F. Amancio, S. Hacon, and C. Barcellos, "Climate-sensitive diseases in Brazil and the world: systematic review (Enfermedades sensibles al clima en Brasil y el mundo: revisión sistemática. Revista Panamericana de Salud Publica)", *Pan Amer. J. Pub. Health,* vol. 42, pp. e85-e85, 2018.
[PMID: 31093113]

[11]    H. Holsinger, N. Tucker, S. Regli, K. Studer, V.A. Roberts, S. Collier, E. Hannapel, C. Edens, J.S. Yoder, and K. Rotert, "Characterization of reported legionellosis outbreaks associated with buildings served by public drinking water systems: United States, 2001–2017", *J. Water Health,* vol. 20, no. 4, pp. 702-711, 2022.
[http://dx.doi.org/10.2166/wh.2022.002] [PMID: 35482386]

[12]    K.M. Benedict, H. Reses, M. Vigar, D.M. Roth, V.A. Roberts, M. Mattioli, L.A. Cooley, E.D. Hilborn, T.J. Wade, K.E. Fullerton, J.S. Yoder, and V.R. Hill, "Surveillance for waterborne disease outbreaks associated with drinking water—United States, 2013–2014", *MMWR Morb. Mortal. Wkly. Rep.,* vol. 66, no. 44, pp. 1216-1221, 2017.
[http://dx.doi.org/10.15585/mmwr.mm6644a3] [PMID: 29121003]

[13]    M.C. Angelici, and P. Karanis, "Protozoan waterborne infections in the context of actual climatic changes and extreme weather events", *Encyclopedia of Environmental Health,* vol. 5, no. 1, pp. 391-399, 2019.

[14]    P.W. Barnes, C.E. Williamson, R.M. Lucas, S.A. Robinson, S. Madronich, N.D. Paul, J.F. Bornman, A.F. Bais, B. Sulzberger, S.R. Wilson, A.L. Andrady, R.L. McKenzie, P.J. Neale, A.T. Austin, G.H. Bernhard, K.R. Solomon, R.E. Neale, P.J. Young, M. Norval, L.E. Rhodes, S. Hylander, K.C. Rose, J. Longstreth, P.J. Aucamp, C.L. Ballaré, R.M. Cory, S.D. Flint, F.R. de Gruijl, D.P. Häder, A.M. Heikkilä, M.A.K. Jansen, K.K. Pandey, T.M. Robson, C.A. Sinclair, S.Å. Wängberg, R.C. Worrest, S. Yazar, A.R. Young, and R.G. Zepp, "Ozone depletion, ultraviolet radiation, climate change and prospects for a sustainable future", *Nat. Sustain.,* vol. 2, no. 7, pp. 569-579, 2019.
[http://dx.doi.org/10.1038/s41893-019-0314-2]

[15]    G.F. Huseien, and K.W. Shah, "Potential applications of 5G network technology for climate change control: A scoping review of singapore", *Sustainability (Basel),* vol. 13, no. 17, p. 9720, 2021.
[http://dx.doi.org/10.3390/su13179720]

[16]    N. Abd El-Mawla, M. Badawy, and H. Arafat, "IoT for the failure of climate-change mitigation and adaptation and IIot as a future solution", *World Journal of Environmental Engineering,* vol. 6, no. 1, pp. 7-16, 2019.
[http://dx.doi.org/10.12691/wjee-6-1-2]

[17]    C. Maraveas, and T. Bartzanas, "Application of Internet of Things (IoT) for optimized greenhouse environments", *AgriEngineering,* vol. 3, no. 4, pp. 954-970, 2021.

[http://dx.doi.org/10.3390/agriengineering3040060]

[18]   A. Salam, "Internet of things for environmental sustainability and climate change", In: *Internet of things for sustainable community development.* Springer: Cham, 2020, pp. 33-69.
[http://dx.doi.org/10.1007/978-3-030-35291-2_2]

[19]   N. Hossein Motlagh, M. Mohammadrezaei, J. Hunt, and B. Zakeri, "Internet of things (IoT) and the energy sector", *Energies,* vol. 13, no. 2, p. 494, 2020.
[http://dx.doi.org/10.3390/en13020494]

[20]   J.T. Kelly, K.L. Campbell, E. Gong, and P. Scuffham, "The internet of things: Impact and implications for health care delivery", *J. Med. Internet Res.,* vol. 22, no. 11, p. e20135, 2020.
[http://dx.doi.org/10.2196/20135] [PMID: 33170132]

[21]   M.J. Baucas, P. Spachos, and S. Gregori, "Internet-of-Things devices and assistive technologies for health care: Applications, challenges, and opportunities", *IEEE Signal Process. Mag.,* vol. 38, no. 4, pp. 65-77, 2021.
[http://dx.doi.org/10.1109/MSP.2021.3075929]

[22]   T.M. Ghazal, "Internet of things with artificial intelligence for health care security", *Arab. J. Sci. Eng.,* pp. 1-12, 2021.

[23]   G. Velliyangiri, A. Venkatachalam, M. Ramachandran, A. Kumar, and M. Subramanian, "Internet of Health Things (IoHT) against COVID-19: A review of recent development", *Computational Intelligence for COVID-19 and Future Pandemics,* pp. 267-279, 2022.

[24]   A. Miles, A. Zaslavsky, and C. Browne, "IoT-based decision support system for monitoring and mitigating atmospheric pollution in smart cities", *J. Decisions Sys.,* vol. 27, no. 1, pp. 56-67, 2018.

[25]   M.Z. Mehmood, M. Ahmed, O. Afzal, and M.A. Aslam, "Internet of Things (IoT) and sensors technologies in smart agriculture: Applications, opportunities, and current trends", In: *In: Building Climate Resilience in Agriculture* Springer, Cham., 2022, pp. 339-364.

[26]   R. Shamshiri, F. Kalantari, K. C. Ting, K. R. Thorp, I. A. Hameed, C. Weltzien, and Z. M. Shad, "Advances in greenhouse automation and controlled environment agriculture: A transition to plant factories and urban agriculture", *IJABE,* vol. 11, no. 1, 2018.
[http://dx.doi.org/10.25165/j.ijabe.20181101.3210]

[27]   J.C.F. Van, P.E. Tham, H.R. Lim, K.S. Khoo, J.S. Chang, and P.L. Show, "Integration of Internet-o--Things as sustainable smart farming technology for the rearing of black soldier fly to mitigate food waste", *J. Taiwan Inst. Chem. Eng.,* vol. 137, p. 104235, 2022.
[http://dx.doi.org/10.1016/j.jtice.2022.104235]

[28]   Z. Luo, J. Zhu, T. Sun, Y. Liu, S. Ren, H. Tong, L. Yu, X. Fei, and K. Yin, "Application of the IoT in the food supply chain- From the perspective of carbon mitigation", *Environ. Sci. Technol.,* vol. 56, no. 15, pp. 10567-10576, 2022.
[http://dx.doi.org/10.1021/acs.est.2c02117] [PMID: 35819895]

[29]   S. Ebiesuwa, O.O. Blaise, A. Adesina, U. Richmond, A. Aderonke, and A.H. Rangkuti, "Internet of Things (IoT) approach to combating economic and environmental issues", *J. Theor. Appl. Inf. Technol.,* vol. 100, no. 10, 2022.

[30]   P.D.C. Fiorini, and B.M.R.P. Seles, "Circular economy business for climate change mitigation: The role of digital technologies", In: *Handbook of Climate Change Mitigation and Adaptation.* Springer International Publishing: Cham, 2022, pp. 3873-3894.
[http://dx.doi.org/10.1007/978-3-030-72579-2_171]

[31]   A. Ben Youssef, and A. Zeqiri, "Hospitality industry 4.0 and climate change", *Circular Economy and Sustainability,* pp. 1-21, 2022.

[32]   A.V. Sheveleva, and M.V. Cherevik, Digital technologies in the oil and gas sector and their contribution to UN climate action goal. In *Industry 4.0: Fighting Climate Change in the Economy of the Future, Springer International Publishing*, 2022, pp. 307-315.

[http://dx.doi.org/10.1007/978-3-030-79496-5_28]

[33] T. Staedter, "Sensor network monitors 1,400-km canal [News]", *IEEE Spectr.,* vol. 55, no. 4, pp. 10-11, 2018.
[http://dx.doi.org/10.1109/MSPEC.2018.8322033]

[34] J. Koehler, P. Thomson, and R. Hope, "Pump-priming payments for sustainable water services in rural Africa", *World Dev.,* vol. 74, pp. 397-411, 2015.
[http://dx.doi.org/10.1016/j.worlddev.2015.05.020]

[35] E. Thomas, L.A. Andrés, C. Borja-Vega, G. Sturzenegger, Ed., *Innovations in WASH Impact Measures: Water and Sanitation Measurement Technologies and Practices to Inform the Sustainable Development Goals.* World Bank Publications, 2018.

[36] J. Jo, B. Jo, J. Kim, S. Kim, and W. Han, "Development of an IoT-based indoor air quality monitoring platform", *J. Sens.,* vol. 2020, pp. 1-14, 2020.
[http://dx.doi.org/10.1155/2020/8749764]

[37] F. Villa-Gonzalez, R. Bhattacharyya, and S. Sarma, "Single and bulk identification of plastics in the recycling chain using Chipless RFID tags", *IEEE International Conference on RFID (RFID),* Atlanta, GA, USA, pp. 1-8, 2021.
[http://dx.doi.org/10.1109/RFID52461.2021.9444372]

[38] E. Wang, S. Attard, M. McGlinchey, W. Xiang, B. Philippa, A.L. Linton, and Y. Everingham, *Smarter irrigation scheduling in the sugarcane farming system using the Internet of Things.* Australian Society of Sugar Cane Technologists, 2019.

[39] V.P. Kovalchuk, T.V. Matiash, V.V. Knysh, O.P. Voitovich, and A.V. Kruchenyuk, "Internet of things (IoT) applications using the LoRaWAN protocol for monitoring irrigated land", *Міжвідомчий тематичний науковий збірник "Меліорація і водне господарство",* no. 2, pp. 130-139, 2019.
[http://dx.doi.org/10.31073/mivg201902-187]

[40] J. Herndon, R. Hoisington, and M. Whiteside, "Deadly ultraviolet UV-C and UV-B penetration to earth's surface: human and environmental health implications. Journal of Geography", *Journal of Geography, Environment and Earth Science International,* vol. 14, no. 2, pp. 1-11, 2018.
[http://dx.doi.org/10.9734/JGEESI/2018/40245]

[41] L.K. Ramasamy, and S. Kadry, "Industrial internet of things", In: *Blockchain in the Industrial Internet of Things.* IOP Publishing, 2021.

[42] P. Ferrari, A. Flammini, E. Sisinni, S. Rinaldi, D. Brandão, and M.S. Rocha, "Delay estimation of industrial IoT applications based on messaging protocols", *IEEE Trans. Instrum. Meas.,* vol. 67, no. 9, pp. 2188-2199, 2018.
[http://dx.doi.org/10.1109/TIM.2018.2813798]

[43] D. Ismail, M. Rahman, and A. Saifullah, "Low-power wide-area networks: opportunities, challenges, and directions", *Proceedings of the Workshop Program of the 19th International Conference on Distributed Computing and Networking,* pp. 1-6, 2018.
[http://dx.doi.org/10.1145/3170521.3170529]

[44] A.R. Sadeghi, C. Wachsmann, and M. Waidner, "Security and privacy challenges in industrial internet of things", *52nd ACM/EDAC/IEEE Design Automation Conference (DAC)* 08-12 June 2015,San Francisco, CA, USA, pp. 1-6.
[http://dx.doi.org/10.1145/2744769.2747942]

[45] S. Gudla, B. Padmaja, P. Mishra, B. Sambana, D. Chandramouli, and A. M. Abbas, "Global warming mitigation using an internet of things based plant monitoring system", *MMTC Communications-Frontiers,* vol. 17, no. 2, 2022.

[46] K.M. Penaskovic, X. Zeng, S. Burgin, and N.A. Sowa, "Telehealth: Reducing patients' greenhouse gas emissions at one academic psychiatry department", *Acad. Psychiatry,* vol. 46, no. 5, pp. 569-573, 2022.

[http://dx.doi.org/10.1007/s40596-022-01698-x] [PMID: 35997996]

[47]   J. Trombley, S. Chalupka, and L. Anderko, "Climate change and mental health", *Am. J. Nurs.,* vol. 117, no. 4, pp. 44-52, 2017.
[http://dx.doi.org/10.1097/01.NAJ.0000515232.51795.fa] [PMID: 28333743]

# CHAPTER 6

# Revolutionizing Precision Agriculture Using Artificial Intelligence and Machine Learning

**Jayalakshmi Murugan[1,*], Maharajan Kaliyanandi[1]** and **Carmel Sobia M.[2]**

[1] *Department of Computer Science and Engineering, School of Computing, Kalasalingam Academy of Research and Education, Krishnankoil, Tamilnadu, India*

[2] *Department of Electrical and Electronics Engineering, PSR Engineering College, Sivakasi, Tamilnadu, India*

**Abstract:** Plant disease mechanization in the agricultural discipline is a major source of concern for every country, since the world's population continues to grow at an alarming rate, increasing the need for food. However, due to a scarcity of necessary infrastructure in various parts of the world, it is difficult to identify them quickly in some areas. In the context of the expanded use of technology, it is now feasible to assess the efficiency and accuracy of methods for identifying illnesses in plants and animals. It has recently been discovered that information technology-based tools, technologies, and applications are effective and realistic measures for the improvement of the whole agricultural field, spanning from scientific research to farmer assistance. The integration of expert systems as a strong tool for stakeholders in agricultural production has enormous promise, and it is now being explored. The suggested effort begins with the collection of disease symptoms and environmental factors by agriculture specialists and plant pathologists, who will then analyze the information gathered. The corrective solution is then recommended to the end user by an expert system, which is accessed through a mobile application. Computer application consisting of an expertise base, inference engine, and a user interface is envisaged as the machine of the future. Integrated inside the gadget is a structured expertise base that contains information on the signs and treatments of various ailments. In order to identify and diagnose plant disorders, the machine must first locate and diagnose the condition. It is accomplished by the analysis of the symptoms of illness on the crop's surface. On the basis of the yield and the surrounding environment, this symptom is utilized to identify the illness and give an entirely unique diagnostic solution. The computer will test the plants and their disordered lives inside the database and provide a set of diagnostic levels in accordance with the condition that the plants are suffering from, according to the database. Farmers may easily identify and manipulate plant diseases with the help of the suggested technology, which is supported by a sophisticated expert system.

\* **Corresponding author Jayalakshmi Murugan:** Department of Computer Science and Engineering, School of Computing, Kalasalingam Academy of Research and Education, Krishnankoil, Tamilnadu, India; E-mail: jayalacsmi@gmail.com

**Keywords:** Plant disease, Expert system, Image-based approach, Knowledge base.

## INTRODUCTION

Plant diseases are one of the most serious problems confronting the agricultural sector across the world. Crop diseases account for one-third of crop output losses each year. On a vast acre of land, it is extremely difficult for farmers to precisely measure crop disease and its symptoms with their own eyes. Early diagnosis of leaf disease is critical to preventing widespread spread from one plant to the entire landscape. Illness control procedures might squander money and lead to future plant losses if the disease is not properly identified.

One of the most important areas where Computer Vision has made a difference is in determining the severity of illnesses. Deep learning [DL], a component of Computer Vision, is effective and shows potential in predicting the harshness of illnesses in vegetation and animals [1, 2], and is utilized to categorize illnesses as well as keep away from late disease recognition. Plant illness differs somewhat from those that impact humans. Diseases are similar due to a variety of variables. The study of data in this sector aids in determining how to improve the utilization of cutting-edge technologies. Images of leaves and other plant components can be used to identify plant illnesses [3]. The technique might be used to analyze photographs of humans, proving the presence of illnesses and determining the amount of their devastation. The purpose of this project is to examine how image-based technologies may be utilized to identify illnesses in both plants and animals.

As a result, we will create a mobile application driven by Machine Learning to automate the diagnosis of tomato leaf disease. Our programme will assist farmers in detecting and alerting them to the early stages of leaf disease. In our proposed system, we employ several feature extraction techniques and classifiers to successfully forecast the leaf disease in its early stage. This saves farmers' time and eliminates the use of ineffective fertilizers, reducing water usage [4], which can harm the health of both plants and soil.

## BACKGROUND

Human-machine communication is enabled through machine learning (ML). It also makes machines behave like people and make decisions for them. It is one of the fastest-growing regions in recent years. ML classifies plant diseases. This method is viewed as a big success in the fight against plant diseases. It has also enhanced agricultural production and limits the water consumption [5]. This technology has been refined over the previous three years to include visualization methods [6]. The issues facing the world today can be addressed if illnesses affecting plants and humans are recognized before they spread across large areas. Diverse ML and DL approaches enable specialists to identify plant diseases and their sources [6]. Several obstacles impact the usefulness and accuracy of this technique in detecting certain disorders.

The first issue is the time required to utilize Deep Learning and Machine Learning, and technologies used to diagnose these illnesses are obsolete or based on old data. Concerning segmentation sensitivity [7], as a result, the region of interest must be extremely precise and sensitive. Another issue is a language barrier that impacts how technology is utilized. Another issue is the lack of resources necessary to maintain this technology. Most ML and DL tasks need a lot of possessions to use. Usually, private and public entities support organization that utilizes this skill to identify illnesses in plants that might impact the study and development of technology.

Plants have become important around the globe. Science and technology have long recognized the importance of plants in medicine, energy production, and current efforts to reduce global warming [8]. Many research initiatives have been launched to give scientists the necessary expertise to construct a cutting-edge convolution system that enables picture identification and illness categorization [9]. The background study indicates that experts in this discipline compare pictures of unaffected and ill vegetation by scanning them.

Decision Trees, K-means, Artificial Neural Networks and Support vector machines (SVM) are some of the ML approaches used to diagnose illnesses [10]. The computers may not operate directly with field photographs. It implies that the photos must be coded into information to supply in to the expert system.

Some plants, like cassava, can be protected from common diseases using image-based detection technologies [11]. Difficulty in meeting objectives due to illnesses threatening some crops. Knowledge about illnesses might assist in averting such situations. Using CNN creates a trustworthy platform for illness analysis [12]. For example, many diseases attack the margins or stem of leaves, whereas others infect the entire leaf. The illness categorization may be determined by analysing the photographs of the leaves.

Cassava is a staple meal in Sub-Saharan Africa because it is high in carbs. This region's nutritional value has been lost due to its sensitivity to viral infections. Africa harvested 145 million tones of cassava in 2014 [12]. Today's food control systems aim to increase productivity. Using modern technology to identify and control illness that impacts productivity is a dependable strategy to reduce food insufficiency. Most plants are raw materials for various industries. Low-quality plants contribute to low-quality goods.

## LITERATURE SURVEY

Tomatoes are a key food crop globally, consumed at 20 kilos per person per year, or 15% of total vegetable consumption. To fulfill global tomato demand, researchers must improve crop productivity and early detection of pests, bacterial, and viral illnesses. Several studies have used artificial intelligence to increase tomato plant longevity *via* early disease identification and control. To decrease agricultural losses, improved crop disease detection, monitoring, and forecast are urgently required. In this context, ML-enabled computer vision has great promise for enhancing crop monitoring [1]. Traditional ML and DL (a subset of ML) approaches perform differently depending on the datasets and tasks performed. To classify tomato diseases, Tan *et al* [2] used the Plant Village tomato dataset and ML/DL models. Crop disease detection by optical inspection of plant leaf symptoms is difficult for farmers with limited resources. The performance of such systems degraded dramatically when evaluated on other datasets or in field situations, according to Lawrence C [3]. The progress thus far is encouraging. Then suggestions are provided on the best architectures to use in both traditional and mobile/embedded computing contexts. Fuentes *et al.* [6] diagnosed illnesses from tomato leaf photos using three detector families: Faster R-CNN, R-FCN, and Single Shot Multibox Detector (SSD). With VGG16 on top of FRCNN, these detectors scored the greatest Average Precision of 83 percent with VGG16 on top of FRCNN. To classify illnesses from tomato leaf photos, Durmuş *et al.* [7] used AlexNet and SqueezeNet architectures. Early identification of plant diseases using computer vision and artificial intelligence (AI) can help lessen disease impacts and overcome human monitoring deficiencies. To diagnose tomato illnesses, Chowdhury *et al.* [8] suggested using a new convolutional neural network dubbed EfficientNet on plain and segmented tomato leaf photos. According to Rehan Ullah Khan *et al.* [9], medical science cannot diagnose all infections in time, which is why certain diseases become pandemics. Our goal is to learn more about illnesses and how artificial intelligence can detect them quickly. To diagnose plant diseases, Liu X *et al.* [13] studied visual identification of plant diseases. Plant disease photos include randomly dispersed lesions, various symptoms, and intricate backgrounds, making them difficult to discriminate. We handled plant disease detection by reweighting visual areas and highlighting sick sections. On-going human monitoring is ineffective in detecting plant illnesses [10]. They employed a one-stage detector called CenterNet to identify and categorize plant illnesses. Intensity fluctuations, colour alterations, and variances in leaf shape and size are used to analyze performance. The provided technique is superior in both qualitative and quantitative analysis. Y. Wu *et al.* [14] addressed VAE strategies to improve agricultural leaf disease detection. VAE utilizes unlabeled data to learn unsupervised, then uses labeled data to learn supervised illness detection. The suggested method's generalization effect and identification accuracy are improved over the classification network using just labeled data. The identification accuracy has grown from 56.13 percent to 78.03 percent for few labeled samples, and it has also increased for numerous labeled samples.

Rice Blast (RB) is a very devastating plant disease found in most rice-growing nations. Infected rice seedlings and early tillers may die. The neck blast infection damages the crop nearing maturity, reducing the yield. Collar infection promotes flag leaf disintegration and impairs photosynthesis for grain filling. Infection of the rice plant's nodes reduces panicle output. The fungal pathogen may attack any part of the rice plant, including the roots, but the sort of infection that causes a blast epidemic depends on the crop's stage of growth. Once established, infection can partially or totally disrupt the rice plant's function, depending on the infection level, plant sections, and stage. Long durations of free wetness on plant surfaces, little or no wind, and high relative humidity are all conducive to the establishment of blast disease. Infection requires leaf wetness or free moisture from rain or other sources. 77-82°F is ideal for germination, infection, and lesion development.

World consumption of rice was 493.13 million metric tonnes in 2019–2020, and 486.62 million metric tonnes in 2018–2019 [15]. In the history of rice farming in Asia, RB outbreaks have caused many tragedies. Due to Asia's traditional dependence on rice farming, RB epidemics have historically resulted in food shortages. Shahriar S *et al.* [16] reviewed the rice blast disease in depth. They claim that only pathogenicity research can reveal pathotypes utilizing a collection of diverse rice cultivars containing varying resistance genes. Resistant genes are now available for rice breeding, however, most efforts focused on monogenic resistance [17]. claimed that knowledge about the pathogen causing the disease, its life cycle and development, epidemiology, symptoms, and management plan can help in predicting disease occurrence and making successful management decisions. Various disease management strategies have been implemented, including resistant cultivars. Rice blast control requires integrated disease management and effective agronomy techniques [18]. Liu *et al.* [19] developed a machine learning-based rice blast prediction model based on short-term environmental data. The model used average, maximum, and minimum air temperatures, relative humidity, soil temperature, and solar energy. F-measures were used to assess the created multilayer perceptron, support vector machine, Elman recurrent neural network, and probabilistic neural network. Finally, the factor significance assessment was subjected to a SA. With a huge dataset of 33,026 photos of six rice illnesses, Deng R *et al.* [20] suggested an autonomous detection approach utilizing deep learning. The Ensemble Model was the method's core. It was then confirmed using a second set of photos. In terms of learning rate, precision, recall, and illness recognition accuracy, DenseNet-121, SE-ResNet-50, and ResNeSt-50 were the top sub models. Deng R *et al.* [21] used AlexNet to recognize common rice leaf diseases such as brown spot, bacterial blight and leaf smut with better results than earlier studies. AlexNet is a deep-learning classification method. Hoang VD *et al.* [22] emphasized extrinsic factors for model training to improve recognition accuracy. With deep learning classification for rice leaf disease identification, certain pretrain models, AlexNet and ResNet101, were implemented as standard convolutional neural network (CNN) architecture. A rule-based model (Yoshino) and a neural network (RNN) were compared by Nettleton, D.F. *et al.* [23]. The results showed that the rice leaf infection detection system could accurately detect the most frequent rice illnesses in real-time. M. Anwarul Azim *et al.* [24] presented a model to identify three rice leaf diseases: B. Color thresholds are used to separate diseased regions from the rest of the picture. Affected areas extract color, shape, and texture characteristics. These attributes can describe the pictures' local and global statistics. Among several classification techniques, this model uses extreme gradient boosting decision tree ensemble for its higher performance. This model outperforms prior efforts on the same dataset by 86.58 percent. Another aspect was the reduction of crop time, the rise in plant density and nitrogen fertilizer use, and the extension of irrigated acreage to allow double rice cropping. Some of these records show the detrimental impact of RBl on rice crop output and farmers' lives.

The Indian economy and most people rely on agriculture. Infected plants and crops produce lower quality and quantity. The difficult issue is to precisely diagnose plant leaf disease. The disease targets leaves, stems, and fruits. It's tough to quantify disorders in necked eyes. Today's society needs a precise visual pattern interpretation. To classify tomato illnesses, Kaur *et al.* [11] employed a pre-trained CNN-based architecture named ResNet. Using a CNN-based architecture, Agarwal *et al.* [12] classified eleven illnesses from tomato leaf photos with 91.2 percent accuracy. Widespread disease prevalence in tomato crops lowers quality and quantity. Early disease detection utilising a proven nondestructive technology can help farmers combat the issue. The authors examined the impact of minibatch size, weight, and bias learning rate on classification accuracy and execution time [25]. Thanjai Vadivel and R. Suguna [26] proposed using CNN (Convolution Neural Network) with computer technology to forecast leaf illness in tomato plants.

Rice is a staple food crop for a substantial portion of the world's population. Rice blast is by far the most serious of the rice illnesses. It's everywhere, it's vital, and it's a threat. Rice blast infections have caused complete rice crop failures. Bombay, Orissa, Kashmir and Kerala have all reported serious losses due to the blast sickness. MacRae put the damage as 69% in the first reported bomb in India in Tanjore in 1948. This is despite the presence of a ruthless and persistent infection. All research-derived plant disease control tactics and procedures have been applied to rice blast, although with mixed results [27]. Even after years of good management, a simple change in the way rice is farmed or the way resistance genes are deployed can result in severe disease losses.

For farmers and agricultural organizations to detect rice leaf illnesses on the go, Gugan Kathiresan *et al.* [28] suggested a high accuracy transfer learning model. It also uses an adversarial generative network to balance disease samples. They also compared it to other transfer learning systems. Tested on a GAN enhanced dataset, the model has an average cross validation accuracy. They used a CNN to identify illnesses and pests in rice plant photos [29]. Rathore *et al.* created a sequential convolutional neural network that automatically extracts information and classifies images. N.S. Patil (2021) [30]. said that detecting agricultural plant diseases using image processing and machine learning approaches can help farmers safeguard paddy crop yields. Feature extraction techniques are used to extract relevant information from the processed pictures. Sethy, P. K. *et al.* [31] examined research publications from 2007 to 2018 with an emphasis on state-of-the-art development. On picture segmentation, feature extraction, feature selection, and classification are compared. They also discuss present progress, limits, and future research directions in rice plant disease diagnostics. In biology, large amounts of pictures are created from single studies and can be utilised for categorization. So biologists must study and extract certain materials to classify. Here, image processing plays a vital role. Diseases caused by climate change destroy considerable crop yields. The most prevalent illnesses are caused by fungi, bacteria, and viruses. Thus, early illness diagnosis is critical. The farmer must constantly examine his harvest with an expert method. So automated and low-cost disease detection is critical in agriculture. Most tomato leaves have fungus, bacteria, and viruses.

## DATA SETS

The inputs utilized in the investigation describe the leaves before and after illness. The data consists of tables and photos of field leaves. The data is examined and categorized in a simple to comprehend manner. Atila *et al.* [32] illustrated the leaves used to identify diseased soybean plants. Fig. (1) shows apparent differences between diseased and unaffected leaves.

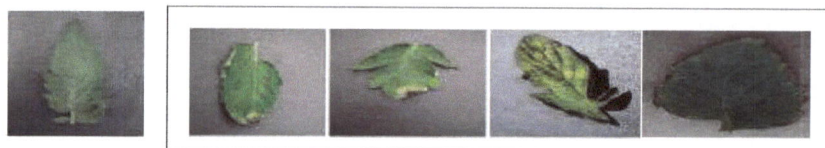

**Fig. (1).** Healthy and diseased Tomato Leaf.

The data set was straightforward to comprehend. The table also indicated the amount of leaves categorized under each illness. There are four categories of leaves investigated in this data collection. All of them have one thing in common: they are easy to analyze and comprehend. Some statistics are also classified according to tiers. Some research findings also indicate the utilization of expertise and its efficacy. We can articulate that inputs recorded by an expert system reveal the sort of control utilized and the efficiency. The data sets assist in understanding how this technology is used and how it affects study quality. The same data displays the categorization methods utilized and their explanations. The data sets for leaf analysis are also derived from field data. The data is reliable since it is based on visible leaf characteristics. The data sets are very straightforward to grasp. According to Ubbens *et al.* [33], the job is divided into illnesses such as bacterial leaf blight and rice blast. Ubbens *et al.* [33] used a PlantVillage data collection in their research. The data collection contains 55,425 photos of 9 crops and 28 plant diseases. The data collection comprised photos of leaves of various hues. Fig. (**2**) shows various PlantVillage data samples.

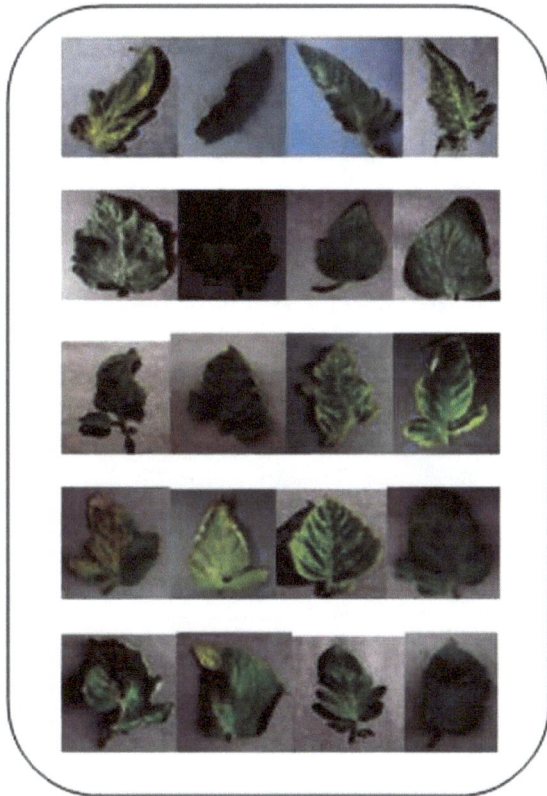

**Fig. (2).** Plant Village Dataset.

The colours represent the diseased leaf portions. The research employed the ImageNet data collection, and the synergy of diverse methodologies resulted in high-quality research outputs. The research data sets utilised in investigations rely on the information contained. For example, Khan *et al.* [34] studied the impact of unmanaged pests in China on overall food production. The research uses vast data sets to illustrate the effects of pests on various productivity levels. Using publicly available data also enabled researchers to understand how their study was conducted and confirmed. The PlantVillage data collection was also utilized by Khan *et al.* to assess 14 distinct cucumber leaf diseases. The data sets were combined to portray the obtained data well. Some authors [35] utilised the PlantVillage data set to illustrate how the data was useful in understanding plant diseases.

Most references utilized data from many researchers aggregated into one set. The research's purpose also influences the data sets' utilization. Using the Plant Village dataset, for example, collects leaves into a table. There is a need to illustrate how illnesses that harm coffee leaves and tomatoes may be diagnosed using image-based detection approaches. Guo *et al.* [36] employed massive numerical and colour data sets to present obtained information. ML algorithm uses the trained input to forecast as well as evaluate new statistics. Data-sets contain anticipated information, to use them in research and establish their validity. The aforementioned data sets have various constraints that limit their application. Northern Leaf Blight (NLB) is another well-known data set utilized by researchers. The photographs were taken using a handheld camera in the field of corn. Unlike the Plant Village data collection, NLB photos were acquired in uncontrolled settings. Using the Plant Village data set, for example, combining data from multiple fields may be difficult. Most data may be varied. The lack of homogeneity may complicate the presentation of data in tables and other displays [37]. The information in the data sets might be better understood if shown graphically [38]. Graphs help readers understand the facts better. Charts can be generated automatically from data sets that contain tables. Suppose in the Plant Village, the overall value is summarized at the conclusion of the table. This information is missing from some data sets. The data sets may also not capture full information about the causes that led to the numbers presented. The data sets may also be difficult to use without prior expertise and knowledge in numerical examination. The data sets also have certain particulars to contradict the results. Sometimes the data sets may hold data regarding illness categorization but not on disease preventative strategies. In certain cases, the data sets are vague, confusing the interpreters. Some statistics may indicate information about illnesses that affect coffee or rice productivity but not the specific impact. This investigation should capture the exact influence on product quantity and quality.

## FEATURE EXTRACTION FOR DISEASE IDENTIFICATION

A knowledge repository, inference engine, and user interface make up an Expert System. A decision support module with interactive user interfaces for diagnostics is part of the proposed expert system. Tomato (Yellow leaf curl virus, Late Blight, Septorial Spot disease) and rice (Blast) plants are included in the system's organised knowledge base, which includes information on disease signs and cures. With an image database connected, the system's decision support becomes more dynamic and user-friendly. The picture database contains images of illness symptoms. The system's goal is to identify and diagnose illness in crops using a variety of methods. This is accomplished by examining the crop for signs of illness. Disease may be identified and treated using this sign on the crop and the surrounding environment. Authorized users will be able to submit information about any crop for disease diagnosis after authenticating the mobile application. The database will be checked to see if the crops and diseases associated with them exist, and the system will then produce a set of diagnostic measures for each one. A plant disease is defined as "anything that prevents a plant from performing to its maximal potential". The existence of a disease caused by a biotic agent absolutely requires the interaction of the following three factors:

• Susceptible host
• A dangerous pathogen and
• An environment favourable for disease development

To find solution, the following process has been proposed. Steps involved in the classification of disease is shown in Fig. (**3**).

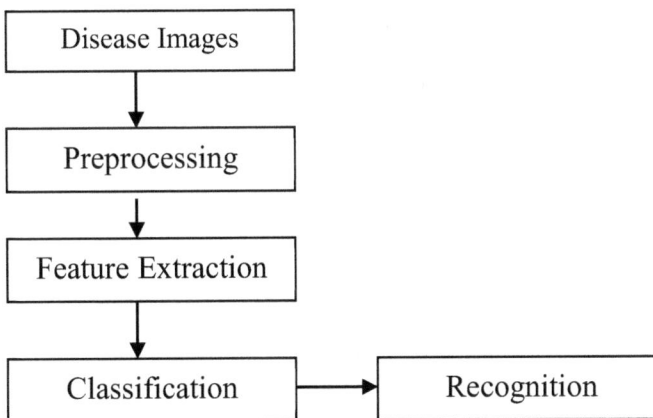

**Fig. (3).** Processing of Feature Extraction and Recognition.

## PERFORMANCE COMPARISON

Our present study involves AI-based plant disease detection in the previous five years. In the following lines, we outline key findings from this research: (i) Plant infection leaf photos are tough to get. As a result, the accessible plant data sets are minimal. We feel that real-world photos are required to make algorithms more realistic. The necessity of the hour is efficient leaf image capture. The research community would appreciate such databases if they were collected in real-time. Images taken using smartphones are more common in recent publications. Moving figure catching systems to the expert system will assist in solving severe database size challenges. (ii) Problems with Feature Extraction Methods Preprocessing, feature extraction, and segmentation are important steps in constructing a machine learning algorithm. Choosing the best preprocessing and segmentation method also relies on the data set. A technique best appropriate for an acquisition typically serves the objective. The reported algorithms have a wide range of spans between modules. Various feature extraction algorithms provide comparable results. In a nutshell, the described procedures need to be standardized. (iii) Classification Module Issues: Plant disease automation and detection have long been a research focus. Researchers find excellent outcomes with few photos for training and testing. This field's researchers examine many classifiers. Based on this research, back propagation neural networks (BPNN), SVM, and linear discriminant analysis (LDA) outperform others A multilayer perceptron follows Naive Bayes, random forest, and knearest neighbor. However, recently released optimized deep neural networks have greatly improved outcomes. Using deep convolutional neural networks properly can assist in improving outcomes for massive data sets. (iv) System Restrictions: We argue that image analysis approaches outperform strategies that visually evaluate illness severity. But these imaging-based solutions aren't ideal. Training data quality directly affects system performance. The performance of a plant disease automation system is heavily influenced by the training photos and extracted characteristics. If these criteria are not met, the system may produce erroneous findings, leading to incorrect illness identification. Researchers must consider adaptive systems with broader needs. Also, generalized strategies that function in varied situations should be adapted. A detailed understanding of processes and tools is also required to improve efficiency. v) Evaluation Tools: There are several ways to compare illness categorization models. True-Positive (TP) indicates the number of infected models successfully detected; True-Negative (TN) indicates the number of correctly recognized unaffected pictures. False-Positive (FP) reflects the number of healthy samples misclassified as infectious. Finally, False-Negative (FN) samples are contaminated samples misclassified as healthy. For example, accuracy is the number of accurate classifications (TP + FP) divided by the total number. Precision is the ratio of properly recognized infected samples

(TP) to all infected samples (sum of TP and FP). A similar ratio, remember, is TP/actual infected samples (sum of TP and FN). F-measure is the mean of accuracy and recall. vi) Results and Comparison: Table **1** compares and summarizes current results for plant disease identification across data sets and techniques. In the following lines, we summarize our findings:

**Table 1. Comparison table.**

| Paper | Year | Dataset | No of Images | Accuracy |
|-------|------|---------|--------------|----------|
| Sembiring *et al.* [39] | 2021 | Plant Village | 54,305 | 0.972 |
| Rangarajan *et al.* [40] | 2018 | Plant Village | 13,262 | 0.950 |
| Adit *et al.* [41] | 2020 | Plant Village | 76,000 | 0.980 |
| Khan *et al.* [34] | 2020 | Plant Village | 7,733 | 0.978 |
| Reddy *et al.* [35] | 2021 | Plant Village | 54,305 | 0.900 |
| Saleem *et al.* [42] | 2020 | Plant Village | 61,486 | 0.730 |
| Stewart *et al.* [38] | 2019 | Aerial image data set | 3,000 | 0.780 |
| Verma *et al.* [43] | 2019 | Plant village | 1520 | 0.952 |
| Ubbens *et al.* [33] | 2018 | Plant Village | 18,160 | 0.962 |
| Lin *et al.* [44] | 2019 | Powdery mildew data set | 20 | 0.961 |
| Atila *et al.* [32] | 2021 | Plant Village | 61,486 | 0.984 |
| Shrivastava *et al.* [37] | 2019 | Oryza sativa database | 619 | 0.914 |
| Tian *et al.* [45] | 2019 | Disease image data set | 640 | 0.94 |

(a) The Plant Village data set served as a testing ground for their hypotheses. Strawberry had the best accuracy (0.991) of all the fruits and vegetables tested. A further benefit of using many classifiers was that the final findings were more accurate. Strawberry also performed well in terms of memory (0.959). The fusion was shown to have the same effect on recall as it did on accuracy. As predicted, strawberry had the best F-measure (0.975). The findings of the F-measure were also improved when the classifiers were fused together. (b) Single Shot Multibox Detector (SSD), Faster Region-based CNN (Faster-RCNN), and Region-based Fully Convolutional Networks were developed by Saleem *et al.* [42] to identify various plant diseases (RFCN). Smaller convolution filters, such as 4x4 and 8x8, were employed in SSD's calculations. Like SSD, object detection is performed in two steps in Faster-RCNN, unlike in SSD. First, the photos are analyzed using feature extractors to generate region recommendations in the region proposal stage. These properties assist to determine which intermediate convolutional layers should have class-specific suggestions. An intermediate picture layer's features are discovered later in the second stage. After ROI pooling, RFCN is quite similar to Faster-RCNN, although it does not include completely convolutional layers. Stochastic Gradient Descent (SGD) was utilized to train a softmax layer by Ubbens *et al*. [33]. After 60 epochs, the convergence was attained. With Bayesian learning techniques, SGD was able to improve the accuracy and F1-score of VGG16, a state-of-the-art image classifier. Both accuracy and an F1-score of 0.94 were achieved using the Monte Carlo (MC) dropout. In addition, the Stochastic gradient Langevin dynamics (SGLD) acquired an accuracy of 0.89, whereas an F1-score of 0.88 was obtained. For out-of-sample classification, SGD has an accuracy of 0.49. As for Monte Carlo (MC) dropout, an accuracy of 0.55 was achieved. Note that the MC dropout can be employed during training and testing. Out-of-sample classification yielded an F1-score of 0.15 for SGLD, compared to 0.54 for in-sample classification. For plant and leaf disease identification, numerous articles applying DL approaches show outstanding results, such as an accuracy better than 0.95 or an F1-score larger than 0.92, according to Lin *et al*. [44]. Atila [32] compared the leaves sickness using AlexNet and GoogleNext and obtained an accuracy of 0.99. Like Atila [32], the Plant Village data set was also used by Rangarajan *et al* [34]. using Machine Learning Techniques and obtained an accuracy of 0.956. Khan *et al*. [36] and Adit *et al*. [41] proposed a CNN-based model and obtained more than 97% accuracy. Reddy *et al*. [35] obtained an F1-score of 0.99 by applying the CNN method, and Sembiring *et al*. [39] developed a concise CNN for detecting the diseases in the tomato plant and obtained an accuracy of 0.972. Table **1** provides a comparison of different approaches.

## CONCLUSION AND FUTURE WORK

In this research, we discussed how ML and DL have helped identify plant diseases. Incorrectly detected illnesses reduce food yields, causing long-term consequences, including global warming and even famine. Multiple works on plant disease automation and detection using ML algorithms are summarized. As shown in the proposed publication, this domain accepts a wide range of CV approaches, making it ripe for further investigation. Several points are given here, which may assist in improving the existing state-of-the-art and provide researchers with some future study suggestions. (i) Identifying Disease Stages: An important aspect of plant disease detection is determining the disease stage. Every illness has phases. Most researchers focused on illness type identification, but none on disease stage identification. These systems must also be able to recommend measures based on the illness stage. Identifying disease forecasts will assist farmers lower the damage percentage. (ii) Disease quantification of a disease is another intriguing subject to investigate. Despite extensive investigation, few scientists have determined the disease's full amount of harm. They can assist a lot because treatment options depend on the disease severity.

This measurement will detect the disease-infected fraction of a culture. This scientific viewpoint is crucial in reducing pesticide use. Most farmers use pesticides to treat illnesses without previous testing or quantification. This is incredibly damaging to human health. Developing an image processing programme will assist in establishing the necessity for certain chemicals. (iii) Mobile and web apps: Several uses of illness identification have been documented. The public can access just a handful of the sites and mobile apps online. Some of these programmes, such as Assess Software and Leaf Doctor, are free to use. But these programmes only operate on photographs with a flat, black backdrop. Thus, online methods and apps for plant disease detection are significantly required. This programme will assist farmers in identifying certain diseases. Such software can generate analysis reports that can be emailed to a disease expert for advice. Examining Transfer Learning to Expand Data Size A similar tendency in CV development towards DL approaches for plant disease diagnosis is not particularly satisfying. Given the complexity of the data, particularly the training stage, transfer learning is the best choice. A heterogeneous domain technique can be used to study knowledge transmission. Keywords for autonomous plant disease detection include LSTMs, optical flow frames, temporal pooling, and 3D convolution. Last but not least, better and more thoroughly designed approaches are required to further explore this subject. For example, data augmentation may be examined further.

## REFERENCES

[1]     A.A. Ahmed, and G.H. Reddy, "A mobile-based system for detecting plant leaf diseases using deep learning", *AgriEngineering,* vol. 3, no. 3, pp. 478-493, 2021.
[http://dx.doi.org/10.3390/agriengineering3030032]

[2]     L. Tan, J. Lu, and H. Jiang, "Tomato leaf diseases classification based on leaf images: A comparison between classical machine learning and deep learning methods", *AgriEngineering,* vol. 3, no. 3, pp. 542-558, 2021.
[http://dx.doi.org/10.3390/agriengineering3030035]

[3]     L.C. Ngugi, M. Abelwahab, and M. Abo-Zahhad, "Recent advances in image processing techniques for automated leaf pest and disease recognition – A review", *Inf. Process. Agric.,* vol. 8, no. 1, pp. 27-51, 2021.
[http://dx.doi.org/10.1016/j.inpa.2020.04.004]

[4]     M. Jayalakshmi, and V. Gomathi, "Sensor-cloud based precision agriculture approach for intelligent water management", *Int. J. Plant Prod.,* vol. 14, no. 2, pp. 177-186, 2020.
[http://dx.doi.org/10.1007/s42106-019-00077-1]

[5]     K. Pavithra, and M. Jayalakshmi, "Analysis of precision agriculture based on random forest algorithm by using sensor networks", *2020 International Conference on Inventive Computation Technologies (ICICT),* Coimbatore, India, pp. 496-499, 2020.
[http://dx.doi.org/10.1109/ICICT48043.2020.9112407]

[6]     A. Fuentes, S. Yoon, S. Kim, and D. Park, "A robust deep-learning-based detector for real-time tomato plant diseases and pests recognition", *Sensors,* vol. 17, no. 9, p. 2022, 2017.
[http://dx.doi.org/10.3390/s17092022] [PMID: 28869539]

[7]    H. Durmus, E.O. Gunes, and M. Kirci, Disease detection on the leaves of the tomato plants by using deep learning, 6th Int.*Conf.* Agro-Geoinformatics, Agro-Geoinformatics, 2017.

[8]    M.E.H. Chowdhury, T. Rahman, A. Khandakar, M.A. Ayari, A.U. Khan, M.S. Khan, N. Al-Emadi, M.B.I. Reaz, M.T. Islam, and S.H.M. Ali, "Automatic and reliable leaf disease detection using deep learning techniques", *AgriEngineering,* vol. 3, no. 2, pp. 294-312, 2021.
[http://dx.doi.org/10.3390/agriengineering3020020]

[9]    R.U. Khan, K. Khan, W. Albattah, and A.M. Qamar, "Image-based detection of plant diseases: from classical machine learning to deep learning journey", *Wirel. Commun. Mob. Comput.,* vol. 2021, pp. 1-13, 2021.
[http://dx.doi.org/10.1155/2021/5541859]

[10]   M. E. Chowdhury, T. Rahman, A. Khandakar, N. Ibtehaz, A. U. Khan, M. S. Khan, and S. H. M. Ali, "Tomato leaf diseases detection using deep learning technique", In: *Technology in agriculture,* 2021.
[http://dx.doi.org/10.5772/intechopen.97319]

[11]   M. Kaur, and R. Bhatia, "Development of an improved tomato leaf disease detection and classification method", *2019 IEEE Conference on Information and Communication Technology* Allahabad, India, pp. 1-5, 2019.
[http://dx.doi.org/10.1109/CICT48419.2019.9066230]

[12]   M. Agarwal, A. Singh, S. Arjaria, A. Sinha, and S. Gupta, "ToLeD: Tomato leaf disease detection using convolution neural network", *Procedia Comput. Sci.,* vol. 167, pp. 293-301, 2020.
[http://dx.doi.org/10.1016/j.procs.2020.03.225]

[13]   X. Liu, W. Min, S. Mei, L. Wang, and S. Jiang, "Plant disease recognition: A large-scale benchmark dataset and a visual region and loss reweighting approach", *IEEE Trans. Image Process.,* vol. 30, pp. 2003-2015, 2021.
[http://dx.doi.org/10.1109/TIP.2021.3049334] [PMID: 33444137]

[14]   Y. Wu, L. Xu, and E. D. Goodman, "Tomato leaf disease identification and detection based on deep convolutional neural network", *Intelligent Automation & Soft Computing,* pp. 561-576, 2021.
[http://dx.doi.org/10.32604/iasc.2021.016415]

[15]   M. Shahbandeh, *Total global rice consumption 2008–2020.* Statisca, 2021.

[16]   S.A. Shahriar, A.A. Imtiaz, M.B. Hossain, A. Husna, and M.N.K. Eaty, "Review: Rice Blast Disease", *Annu. Res. Rev. Biol.,* pp. 50-64, 2020.
[http://dx.doi.org/10.9734/arrb/2020/v35i130180]

[17]   G. O. Agbowuro, M. S. Afolabi, E. F. Olamiriki, and S. O. Awoyemi, "Rice blast disease (Magnaporthe oryzae): A menace to rice production and humanity", *Intern. J. Patho. Res.,* pp. 32-39, 2020.

[18]   W. Liang, H. Zhang, G. Zhang, and H. Cao, "Rice blast disease recognition using a deep convolutional neural network", *Sci. Rep.,* vol. 9, no. 1, p. 2869, 2019.
[http://dx.doi.org/10.1038/s41598-019-38966-0] [PMID: 30814523]

[19]   L.W. Liu, S.H. Hsieh, S.J. Lin, Y.M. Wang, and W.S. Lin, "Rice blast (magnaporthe oryzae) occurrence prediction and the key factor sensitivity analysis by machine learning", *Agronomy,* vol. 11, no. 4, p. 771, 2021.
[http://dx.doi.org/10.3390/agronomy11040771]

[20]   R. Deng, M. Tao, H. Xing, X. Yang, C. Liu, K. Liao, and L. Qi, "Automatic diagnosis of rice diseases using deep learning", *Front. Plant Sci.,* vol. 12, p. 701038, 2021.
[http://dx.doi.org/10.3389/fpls.2021.701038] [PMID: 34490004]

[21]   M.M.H. Matin, A. Khatun, M.G. Moazzam, and M.S. Uddin, "An efficient disease detection technique of rice leaf using alexNet", *Journal of Computer and Communications,* vol. 8, no. 12, pp. 49-57, 2020.
[http://dx.doi.org/10.4236/jcc.2020.812005]

[22]    V.D. Hoang, "Rice leaf diseases recognition based on deep learning and hyperparameters customization", In: *International Workshop on Frontiers of Computer Vision* Springer: Cham, 2021, pp. 189-200.
[http://dx.doi.org/10.1007/978-3-030-81638-4_16]

[23]    D.F. Nettleton, D. Katsantonis, A. Kalaitzidis, N. Sarafijanovic-Djukic, P. Puigdollers, and R. Confalonieri, "Predicting rice blast disease: Machine learning versus process-based models", *BMC Bioinformatics,* vol. 20, no. 1, p. 514, 2019.
[http://dx.doi.org/10.1186/s12859-019-3065-1] [PMID: 31640541]

[24]    M.A. Azim, M.K. Islam, M.M. Rahman, and F. Jahan, "An effective feature extraction method for rice leaf disease classification", *TELKOMNIKA (Telecommunication Computing Electronics and Control),* vol. 19, no. 2, pp. 463-470, 2021.
[http://dx.doi.org/10.12928/telkomnika.v19i2.16488]

[25]    A.K. Rangarajan, R. Purushothaman, and A. Ramesh, "Tomato crop disease classification using pre-trained deep learning algorithm", *Procedia Comput. Sci.,* vol. 133, pp. 1040-1047, 2018.
[http://dx.doi.org/10.1016/j.procs.2018.07.070]

[26]    T. Vadivel, and R. Suguna, "Automatic recognition of tomato leaf disease using fast enhanced learning with image processing", *Acta Agric. Scand. B Soil Plant Sci.,* pp. 1-13, 2021.

[27]    R. Bhagwat, and Y. Dandawate, "A review on advances in automated plant disease detection", *Int. j. eng. technol. innov.,* vol. 11, no. 4, pp. 251-264, 2021.
[http://dx.doi.org/10.46604/ijeti.2021.8244]

[28]    G. Kathiresan, M. Anirudh, M. Nagharjun, and R. Karthik, "Disease detection in rice leaves using transfer learning techniques", *J. Phys. Conf. Ser.,* vol. 1911, no. 1, p. 012004, 2021.
[http://dx.doi.org/10.1088/1742-6596/1911/1/012004]

[29]    N.P.S. Rathore, and L. Prasad, "Automatic rice plant disease recognition and identification using convolutional neural network", *J. of Critical Rev.,* vol. 7, no. 15, pp. 6076-6086, 2020.

[30]    N.S. Patil, "Identification of paddy leaf diseases using evolutionary and machine learning methods", *Turk. J. Comput. Math. Edu.,* vol. 12, no. 2, pp. 1672-1686, 2021.

[31]    P.K. Sethy, N.K. Barpanda, A.K. Rath, and S.K. Behera, "Image processing techniques for diagnosing rice plant disease: A survey", *Procedia Comput. Sci.,* vol. 167, pp. 516-530, 2020.
[http://dx.doi.org/10.1016/j.procs.2020.03.308]

[32]    Ü. Atila, M. Uçar, K. Akyol, and E. Uçar, "Plant leaf disease classification using EfficientNet deep learning model", *Ecol. Inform.,* vol. 61, p. 101182, 2021.
[http://dx.doi.org/10.1016/j.ecoinf.2020.101182]

[33]    J.R. Ubbens, and I. Stavness, "Corrigendum: Deep plant phenomics: A deep learning platform for complex plant phenotyping tasks", *Front. Plant Sci.,* vol. 8, p. 2245, 2018.
[http://dx.doi.org/10.3389/fpls.2017.02245] [PMID: 29375612]

[34]    M.A. khan, T. Akram, M. Sharif, and T. Saba, "Fruits diseases classification: Exploiting a hierarchical framework for deep features fusion and selection", *Multimedia Tools Appl.,* vol. 79, no. 35-36, pp. 25763-25783, 2020.
[http://dx.doi.org/10.1007/s11042-020-09244-3]

[35]    S.R.G. Reddy, G.P.S. Varma, and R.L. Davuluri, "Optimized convolutional neural network model for plant species identi fi - cation from leaf images using computer vision", *Int. J. Speech Technol.,* vol. 26, no. 1, pp. 23-50, 2021.

[36]    Y. Guo, J. Zhang, and C. Yin, "Plant disease identification based on deep learning algorithm in smart farming", *Discrete Dynamics in Nature and Society,* vol. 2020, p. 11, 2020.
[http://dx.doi.org/10.1155/2020/2479172]

[37]    V.K. Shrivastava, M.K. Pradhan, S. Minz, and M.P. Thakur, ""Rice Plant disease classification using

transfer learning of deep convolution neural network," ISPRS -", *Int. Arch. Photogramm. Remote Sens. Spat. Inf. Sci,* vol. XLII-3, no. W6, pp. 631-635, 2019.
[http://dx.doi.org/10.5194/isprs-archives-XLII-3-W6-631-2019]

[38]    E.L. Stewart, T. Wiesner-Hanks, N. Kaczmar, C. DeChant, H. Wu, H. Lipson, R.J. Nelson, and M.A. Gore, "Quantitative phenotyping of Northern Leaf Blight in UAV images using deep learning", *Remote Sens.,* vol. 11, no. 19, p. 2209, 2019.
[http://dx.doi.org/10.3390/rs11192209]

[39]    A. Sembiring, Y. Away, F. Arnia, and R. Muharar, "Development of concise convolutional neural network for tomato plant disease classi fication based on leaf images", *J. Phys. Conf. Ser.,* vol. 1845, no. 1, p. 012009, 2021.
[http://dx.doi.org/10.1088/1742-6596/1845/1/012009]

[40]    T.A. Salih, A.J. Ali, and M.N. Ahmed, "Deep Learning Convolution Neural Network to Detect and Classify Tomato Plant Leaf Diseases", *Open Access Library Journal,* vol. 7, p. e6296, 2020.
[http://dx.doi.org/10.4236/oalib.1106296]

[41]    V.V. Adit, C.V. Rubesh, S.S. Bharathi, G. Santhiya, and R. Anuradha, *A Comparison of deep learning algorithms for plant disease classification," in advances in cybernetics, cognition, and machine learning for communication technologies, lecture notes in electrical engineering.* vol. 643. Springer: Singapore, 2020, pp. 153-161.

[42]    M.H. Saleem, S. Khanchi, J. Potgieter, and K.M. Arif, "Imagebased plant disease identification by deep learning meta-architectures", *Plants,* vol. 9, no. 11, p. 1451, 2020.
[http://dx.doi.org/10.3390/plants9111451] [PMID: 33121188]

[43]    S. Verma, A. Chug, A.P. Singh, S. Sharma, and P. Rajvanshi, *Deep Learning-based mobile application for plant disease diagnosis: A proof of concept with a case study on tomato plant, in applications of image processing and soft computing systems in agriculture.* IGI Global, 2019, pp. 242-271.
[http://dx.doi.org/10.4018/978-1-5225-8027-0.ch010]

[44]    K. Lin, L. Gong, Y. Huang, C. Liu, and J. Pan, "Deep learningbased segmentation and quantification of cucumber powdery mildew using convolutional neural network", *Front. Plant Sci.,* vol. 10, p. 155, 2019.
[http://dx.doi.org/10.3389/fpls.2019.00155] [PMID: 30891048]

[45]    Y. Tian, G. Yang, Z. Wang, E. Li, and Z. Liang, "Detection of apple lesions in orchards based on deep learning methods of cyclegan and yolov3-dense", *Journal of Sensors,* vol. 2019, p. 13, 2019.
[http://dx.doi.org/10.1155/2019/7630926]

# CHAPTER 7

# Internet of Fisheries Things (IOFT) for Blue Economy & Ecosystem

**Sadiq Mohammed Sanusi**[1,*], **Singh Invinder Paul**[2] and **Ahmad Muhammad Makarfi**[3]

[1] *Department Of Agricultural Economics & Extension, Federal University Dutse, P.M.B. 7156, Dutse, Nigeria*

[2] *Department of Agricultural Economics, SKRAU, Bikaner, India*

[3] *Department of Agricultural Economics and Extension, BUK, Kano, Nigeria*

**Abstract:** The industry with the fastest global growth is aquaculture, particularly in developing nations. Studies of control systems, automation, IoT, and artificial intelligence are becoming more and more prevalent as a result of developments in information technology and embedded systems in the digitalization era. The 4.0 industrial revolution, which promotes device connection *via* IoT and AI, is a result of the swift advancement of science and technology. One of the key factors in the growth of this revolution is IoT. The fourth industrial revolution is a shift from manual to automated operations that rely on computer technology. The digitalization of all industries, including fisheries and aquaculture, must be the first step in technological advancement toward the usage of the Internet of Things (IoT) and Big Data. The raising and catching of fish have undergone several changes as a result of digitization in the fishing industry. The main benefits of digitizing fisheries are those related to time, money, and labor efficiency. The procedure of digitizing aquaculture can be seen from a variety of angles, including pre-production (locating cultivation sites using GIS), production (improving resource efficiency by employing automatic feeders and water quality testing equipment), and post-production (utilization of digital marketing).

**Keywords:** Digitalization, Ecosystem, Fisheries: Aquaculture, IoT.

## INTRODUCTION

For hundreds of millions of people around the world, fisheries and aquaculture continue to be a significant source of food, nutrition, money, and livelihood [1]. The global fish supply in 2016 reached 171 million tons, of which 88 percent were used for direct human consumption, according to The State of World Fisheries and Aquaculture 2018 [2]. This was made possible by relatively stable capture fisheries production, decreased wastage, and

---

* **Corresponding author Sadiq Mohammed Sanusi:** Department Of Agricultural Economics & Extension, Federal University Dutse, P.M.B. 7156, Dutse, Nigeria; Tel: +2347037690123; E-mail: sadiqsanusi30@gmail.com

ongoing aquaculture growth. One of the agricultural subsectors with the quickest growth is fisheries and aquaculture [3]. It is crucial in providing for the rising population's nutritional and food security. The availability of fish for human consumption has increased more rapidly than the world's population over the last 50 years [4]. India is second among all nations in both aquaculture production and overall fish production [3].

The conservation of oceans, seas, and marine resources while promoting those ecosystem services that are of critical value for humans is one of the main Sustainable Development Goals (SDG) 14, which includes fisheries as a vital component [5]. The benefits of aquatic ecosystems for both current and future generations will be enabled by more sustainable resource use, adjustments to production patterns, and enhanced control of human activities [2]. The advantages that fishing resources offer must be safeguarded through sustainable management and value addition as the population of the world expands and the need for food and jobs rises along with it. Regardless of the type of water available or the species chosen, every fish completely depends on the water to live, eat, develop, and conduct other bodily activities. It follows that it is not surprising that the water quality of a fish farming facility plays a significant role in its performance. Some of the most crucial characteristics in aquaculture are temperature and pH [6]. Monitoring of the water quality is the only method used for fisheries management.

The earth's ecology depends critically on water, a unique type of environmental resource. In addition to sustaining human life and behavior, water serves as an ecosystem that helps protect regional environments, ecological processes, and ecological structure [7]. Although water cannot be replenished, it must still be used. Water contamination is one of the many major environmental issues that are emerging due to the economy's rapid growth. On March 6, 2019, a purportedly illegal Indian tyre recycling facility allegedly dumped chemical trash in Sungai Kim Kim in Pasir Gudang Johor, putting numerous lives at danger [2]. The entire lake's water is contaminated as a result. Thus, thousands of fish are affected. Temperature, pH, and turbidity are regularly checked factors of water quality. It is crucial to assess water quality because a variety of variables, such as pollution, wind, weather, and rain runoff, can have an impact on fish health [8]. The parameters aid in reducing future problems by enabling ongoing monitoring. They must incorporate the Internet of Things (IoT) into every aspect of modern life in order to improve easy and seamless living [9].

An innovative new idea called the Internet of Things (IoT) has the power to make almost everything "smart" [10]. An IoT network consists of real-world items. The

IoT aims to make it possible for objects to connect with anything and anybody at anytime, anywhere, and preferably *via* any network or path [11]. Machine-to-machine communication, wireless sensor networks, WI-FI, GPS, microcontrollers, and processors are only a small portion of the Internet of Things (IoT). IoT is a collection of hardware, software and communications technologies that are used to store, retrieve, and process data. These technologies include electronic systems. Through the current network, it enables remote control or object monitoring for any object from anywhere in the world. The internet of things has benefits and drawbacks, but the fact that there are several obstacles and problems with its implementation must be acknowledged (IoT).

## DIGITALIZATION OF AQUACULTURE

The fourth industrial revolution was characterized by automated processes based on computer technology [12]. With the use of big data, IoT, and CPSs (cyber-physical systems), industrial revolution 4.0 aims to introduce and establish "smart factories," whose operations are enhanced and altered by technology [13, 14]. The phrase "Internet of Things" (IoT) was first used by Kevin Ashton in 1999 to describe the fundamental idea of computing, which enables an object (or thing) to hear, see, think, and operate by using information to make decisions [15]. The digitalization process in all industries, including fisheries, must be the starting point for technological advancements towards IoT and big data.

The way fish are raised and caught has changed significantly as a result of digitization in fisheries. The main benefits of digitizing fisheries are those related to time, money, and labor efficiency. Since the start of the third industrial revolution in 1969, there has been a transition from manual to automatic production processes through the use of electronic tools and information technology [14], as well as the use of GPS (Global Positioning System) or remote sensing systems as satellite-based navigation systems on fishing vessels [16]. The advancement of science and technology has rendered national boundaries obsolete, making it easier than ever before for countries to trade knowledge and even data. Fisheries digitization attempts to reduce illegal, unreported, and unregulated fishing, manage fishing grounds, improve land conditions, map aquaculture areas, and even increase worker productivity.

When it comes to fisheries, where the goal is to maximize the potential of already-existing resources, maritime developing nations with a wealth of fisheries resources must utilize digitization and even participate in technology development. In order to serve as an example of the application of technology, particularly in aquaculture, it is also required to keep a record of the digitalization of fisheries in marine developing nations. In an effort to improve productivity, lessen negative environmental effects, and organize sustainable cultivation activities, technology is being used to pinpoint the location of production using the GIS application. The use of technology is continued in the cultivation process once the culture site has been selected. Examples include the use of automatic feeders and automatic monitoring water systems to boost productivity, improve the feeding conversion ratio, and offer real-time water quality data.

## Definition of Digitization in the Fisheries Sector

Technology and communication advances have had an impact on fisheries activity. The industrial revolution 4.0, according to Dobrzański and [13], has made digitalization a solution. This is true because the agri-food industry, which includes the fisheries sector, must assure the establishment of food security to meet the needs of 9.6 billion people by 2050. In order to make fisheries activities more efficient in terms of economy, labor, and time, various forms of digitization must be used in the sector of fisheries. These include accessing the most recent information, being aware of market developments, calculating input and output, and increasing production.

By employing information technology and computational processes, the Internet of Things (IoT) is a concept that enables numerous things to connect with one another, share data, and even make choices together. The inventor of the MIT Auto-ID Center, Kevin Ashton, coined the phrase "Internet of Things" in 1999. Regarding the definition of IoT, there were three categories: interactions between people and other people online; interactions between people and machines online; and interactions between machines themselves. IoT's goal is to make it possible for any object to connect with anything or anybody across numerous networks at any time and any place.

Big data, according to [14], is a concept that refers to a collection of data that has the characteristics of a high volume, velocity, and variety of data and that needs technology and particular analytical techniques to be transformed into something more valuable. In the early 1990s, there was a significant switch from analog to digital signals that marked the beginning of the digitization process. A data set structure will be created from analog signals that have been converted to digital representation. The current state of big data technology is one of rapid development, driven by three key factors: 1) the quick development of data storage capabilities, 2) the quick development of data processing engine capabilities, and 3) the profusion of data availability. Big data and the internet of things have been extensively used in many industries, including aquaculture.

## Digitalization Use in the Aquaculture Sector

### *Location Determination Using GIS*

Fishery digitalization, as earlier said, aims to boost output and improve resource efficiency (Fig. **1**). Using remote sensing, one may locate suitable fishing spots as well as assess the viability of potential aquaculture sites. The purpose of location determination is to make better use of the available time and resources. According to [14] and [17], because geospatial research was becoming more popular and employed in a wide range of fields of study, there was a significant surge in the development and exploration of GIS systems in the early 1995s. Additionally, the use of GIS in fisheries is an effort to improve the profitability of fishing activities, and some of the data required to assess the viability of production include economic aspects, social aspects, physical or biological aspects, and sustainability factors.

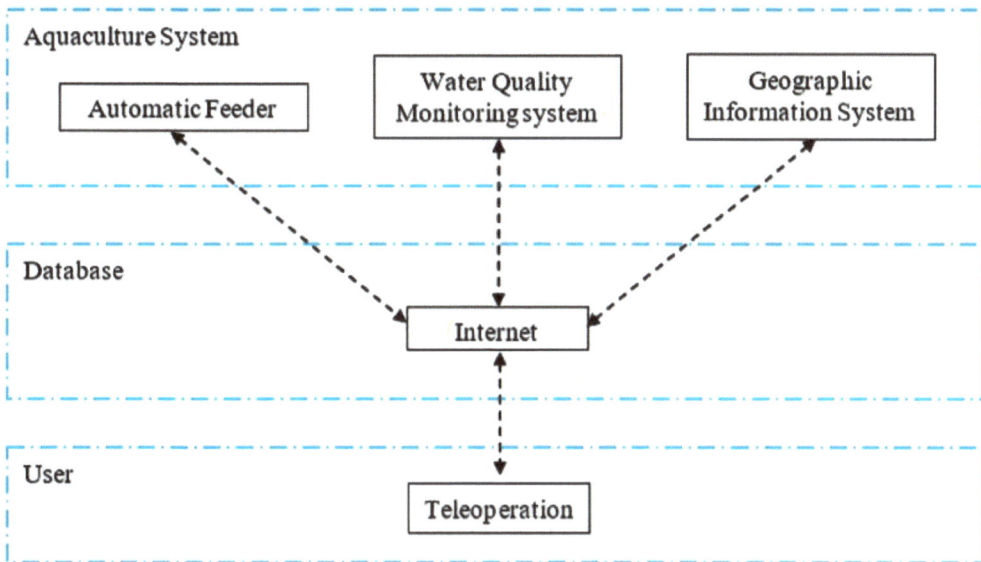

**Fig. (1).** Digitalization use in the aquaculture sector.

Because GIS can analyze, organize, and display a wide variety of spatial data, it has been widely used in many different disciplines of research [18]. The physical, chemical, and biological characteristics that will produce sustainable aquaculture should be taken into consideration while choosing an aquaculture location. To make the findings of data interpretation easier to read and understand, the environmental quality and land suitability data will be displayed on a thematic map. Using GIS to locate aquaculture has the benefit of obtaining more thorough and complete data, which aids in planning aquaculture activities, resource management, and land use.

The choice of a location is vital in aquaculture since the optimum spot will maximize production efficiency and minimize any negative environmental effects. The best location for aquaculture activities would be determined by applying GIS in conjunction with multi-criteria decision making methods (MCDM), as this combination will result in more data and higher-quality final conclusions. Aquaculture activity planning will be more mature, extensive, and long-term with digitization in the pre-cultivation stage.

## *Making Use of Technology for Automatic Feeders*

The availability of feed accounts for 60-70% of production expenses, according to [19] and [20], making it the most crucial factor in determining the success of aquaculture activities. Premised on these various factors, it is imperative to provide feed to aquaculture products as efficiently as possible, taking into account the feed's amount, quality, and manner of appropriation. By developing automatic feeders, technological advancements enabled modernisation and increased efficiency in fish farming operations. A device to automatically feed fish, automatic feeders can be programmed to respond to time, water temperature, phone calls, feeding behavior, self-feeding, auditory feedback and mobile robotics.

A measured feed supply employing an autonomous feeder, according to [14], could suppress externalities from cultivation activities. In order to toss the feed and adjust it based on the schedule that was established before the automatic feeder was deployed, [21] created an automatic feeder mechanism. Additionally, the microcontroller has the ability to alert farmers when the feed storage container is empty. The temperature of the water is used in another study's automatic feeder control. Since fish hunger and growth rate can be influenced by the water temperature, the automatic feeder was created. The results of the microcontroller's analysis can be used to control an automatic feeder that is temperature-dependent to control when and how much feed is fed to crops. A telephone-controlled automatic feeder was created by [22]. When there was an incoming call, the phone set up in the automatic feeder would switch on. This provided an electric voltage to operate the dynamo, allowing feed to automatically fall into the cultivation tank.

To ascertain the feeding behavior of domesticated species, [23] uses a camera utilized for machine vision. The microcontroller will process and analyze the images from the camera to extract behavioral indicators. If the fish in the tank are swimming around and not responding to the meal, the microcontroller will send a command to the feeding machine to stop. The outcome demonstrates that using automatic feeders designed based on fish behavior can lessen feed residues, hence maintaining the impact on water quality and improving feeding efficiency. The autonomous feeder created by [24], sometimes known as the self-feeder, is controlled by fish. A switch with an orange bead that had a diameter of 5 mm was used by the self-feeder and was positioned in the center of the culture container.

The SS-5GL sensor is triggered when the fish tugs the bead, turning the feed storage container 90 degrees so that the feed will fall into the culture container. Fish grown in a self-feeder, both indoors and outdoors, were compared for the purpose of study [14]. The deployment of a self-feeder was proven to be an efficient technique for tilapia farming, but only when the water is warm enough (24-26°C) to preserve the fish's physiological condition. [25] and [26] employed a hydrophone in their studies, in contrast to [23], to control how much food the automatic feeder fed the animals. The automatic feeder is made to employ a hydrophone, a sound detection device submerged in water, to listen for the sound made when fish or shrimp are eating and use it to estimate how much feed is being consumed by aquaculture commodities. As a result of the hydrophone's inability to pick up sound when the aerator is operating, the automatic feeder is only used from 7 AM to 9 AM. According to findings from both researches, acoustic feedback-based automatic feeders significantly outperform schedule-controlled automatic feeders in terms of growth rate and efficacy.

Diverse feed technology advances have been created to boost aquaculture commodity production and lessen the negative effects of unusable feed waste. The majority of modern commercial fish farming makes extensive use of feeding technologies like automatic feeders. By using an automatic feeder, mistakes in manually feeding stocked animals can be prevented. In other words, this technology significantly enhances feed management. Demand or self-feeding feeders are modifications of automatic feeders. Based on the fish's nutrition, this feeding technique makes use of the fish's appetite. Reduce feed conversion, improve growth rate, and lessen uneaten feed are the goals of the automatic feeder. This assertion has been supported by a number of studies, including those on the European Seabass (*Dicentrarchus labrax*), Yellowtail (*Seriola quinqueradiata*), and Tilapia (*Oreochromis niloticus*).

The purpose of the 3.0 industrial revolution is to use robots to simplify human activities. [27] created a robot that performs human responsibilities in feeding in aquaculture activities by utilizing technological advancements. The mobile robot (MR) is a machine that moves along a rail (monorail) on four wheels and has been pre-programmed with information about quantity, pond or tank location, and feeding schedule. Additionally, MR has the ability to operate in supervised mode or automatic mode (using an embedded application) (by utilizing the internet connection to the computer). This feeding robot is being used in an effort to simplify fish farming and research procedures, reduce mistake rates, particularly in research procedures, and boost aquaculture productivity.

## *Automatic Evaluation of Water Quality*

Water quality is a major factor in aquaculture activities' performance, in addition to the availability and high quality of feed. According to [28], preserving water quality is essential for ensuring that aquaculture operations are more productive while also conserving the ecosystem and ecological factors. In aquaculture operations, good water quality will promote fish growth and lower the risk of fish disease. Dissolved oxygen (DO), pH, temperature, ammonia, nitrate, and salinity (for aquaculture activities in brackish-sea waters) are a few characteristics that must be understood to assure the success of aquaculture activities. [14] added that it is important to understand these parameters because they have a significant impact on the success of aquaculture. For instance, temperature has an impact on fish development and appetite, while DO has an impact on stress levels, activity, and mortality. Ammonia toxicity is influenced by pH because anionic ammonia becomes more poisonous at high pH levels (also known as alkaline circumstances).

Since even a small change in the water can have an impact on fish physiology, real-time data is a crucial tool for intensive aquaculture systems. To ensure the success of aquaculture, a real-time automatic water quality monitoring system is required. The ability to assess water quality has been improved *via* the performance of several studies. Technology to manage water quality utilizing network wireless was developed by [28]. Using inexpensive open-source hardware, [29] created a monitoring tool. The IoT-based water quality monitoring system was created by [30]. The outcomes of these researches are identical. Aquaculture activities are made simpler by the installation of an automatic water quality monitoring system since water quality data will be reported continually that will be needed to make a choice. The information is given to help with stage decisions, feeding additions or reductions, fish density changes or absences, and water change decisions. The digitization of water quality control would thereby increase the efficiency and effectiveness of fish farming activities.

## *Marketing Aquaculture Products Online*

The transformation includes both pre- and post-cultivation operations, including the marketing of aquaculture products, as well as smart aquaculture, automatic feeders, and automatic water quality monitoring systems. Pre- and post-cultivation activities include determining aquaculture locations using geographic information systems (GIS). The market is moving to the internet market, also known as e-commerce, as a result of the development of information and technology. The online market has the potential to lower supply chain costs, boost fish farmers' income, and raise the market share of items made from aquaculture. For example, in Indonesia, with more than 100 million social media users, online marketing is the ideal method for distributing aquaculture products there. This presents a significant opportunity for product marketing. South-east Asian maritime nations face a challenge with the marketing of agricultural products that is known for its lengthy supply chain and impacts the selling price of these products. This challenge can be solved by using digital product marketing. Nevertheless, it turns out that online marketing activities also affect consumer behavior in addition to having an impact on a product's supply chain. [31] explained this in their research, 60% of respondents choose to shop for aquaculture products online because it is more convenient, 67% are satisfied with the price range and quality of the products offered through online marketing, and 89% of respondents made a purchase.

## IOT FOR MONITORING SHRIMP/FISH POND

Sensor networks, hardware circuits, energy, and analytic software are only a few of the components that make up an entire Internet of Things system. Data analysis and forecasting techniques are crucial jobs in the IoT system for environmental monitoring. Three Indian businesses were discovered to be engaged in IoT. Building Internet of Things (IoT) systems to measure various water factors that are essential for the growth and survival of shrimps in an aquaculture pond is the focus of Andhra Pradesh's Eruvaka Technologies, a company founded in 2012. Sensor data is uploaded to the cloud and then sent to specific clients *via* an Android app downloaded to their phones. The oxygen content, temperature, and pH range of the pond are among the data gathered. In addition to using the app to obtain information, it can also be sent by SMS or voice call. In other words, the users of this system can view all of the signs on their computers, laptops, or mobile devices. When the value of an indicator reaches the upper limit (threshold) value, the system automatically sends alerts to the users (farmers). The system may also predict each indicator's values for upcoming dates (or any chosen time) based on past data (Fig. **2**).

**Fig. (2).** IoT for monitoring shrimp/fish pond.

Aerators and feeders can be automated across numerous ponds with the help of cFog, an Andhra Pradesh-based company founded in 2017. It also checks metrics for dissolved oxygen, pH, ammonia, and temperature. In 2017, the Chennai-based start-up Aqua Connect from Tamil Nadu created Farm MOJO, a farm advisor that employs machine learning to give shrimp farmers insights about how to manage the quality of the water in their ponds, the information they need to feed their shrimp, and the shrimps' pace of growth [3].

## APPS FOR FISHERIES AND AQUACULTURE USE OF IOT IN MOBILE

Adopting cutting-edge technologies like blockchain, AI, IoT, and mobile apps is necessary to realize the Blue Revolution (Neel Kranti Mission) and transform fisheries into a modern, global industry [3]. The blue revolution may be ushered in large part by these technologies. The IoT and mobile app industries are expanding regularly among the new technologies. Mobile apps are a major source of revenue generating and are becoming more prevalent in the industry [32]. The proper kind of information may be disseminated at the right moment with the use of mobile apps. With 560 million internet subscribers in 2018, India is one of the largest and fastest-growing markets for digital consumers, second only to China, according to the GOI study in 2019 of Digital India-Technology to Transform a Connection Nation [3]. India's average monthly mobile data consumption is 8.3 gigabits (GB), compared to China's mobile users' 5.5 GB and South Korea's sophisticated digital economy's 8-8.5 GB. In 2018, more apps were downloaded by Indians (12.3 billion) than by citizens of any other nation besides China, which had 1.2 billion subscribers to mobile phones. India is digitizing more quickly than any other nation, and more than 40% of people have access to the internet. Connecting common objects to the internet so they can exchange and gather data is what the Internet of Things (IoT) is all about [33]. It is anticipated that 75.44 billion devices would be a part of the IoT by 2025 as asserted by Statista in 2019

[3]. Applications for the Industrial Internet of Things, Smart Cities, Smart Retail, Smart Grids, Smart Homes, Wearables, Connected Cars, Healthcare, Poultry, and Farming are all accessible.

Modern technology facilitates effective resource use and reduces the need for human labor in many areas of daily life [34, 35]. IoT in fisheries and aquaculture has a lot of promise, and fishermen and farmers can gain an advantage in a cutthroat market. Aquaculture-related apps offer data on market prices, news, and events, disease alerts, culture-related information, inputs, and equipment. The potential fishing zone (PFZ), ocean state forecast, wind speed, wave heights, disaster warnings, Global Positioning System (GPS), tsunami warning, and search and rescue information are all provided by marine fisheries-related apps. Marketing-related apps offer home delivery of marine and freshwater fish, shrimp, crabs, lobsters, and other seafood in both raw and dressed forms.

## Aquaculture-Related Mobile Applications

The field of aquaculture is expanding. Contrary to earlier predictions of a decline in productivity, it is anticipated that India would produce a record amount of farmed shrimps this fiscal year. The industry had anticipated a 10-20% decline in sales due to low stock, low prices, and illnesses in the first half of the fiscal year. However, exports started to grow up later, and with a 27% share, India is the main export market for shrimp. After the US (42%), China has overtaken it as the second-largest customer with a 25% stake. The expansion of online and in-store shopping in 2020 will result in continuous growth. It is anticipated that new technologies like IoT would be implemented in this field.

1. *Aqua Brahma*: A 2015 app created by Mile Deep Works Pvt. Ltd. (Fig. **3**). It enables farmers to maintain contact with vendors and hatcheries and to be informed about news, market indicators, and material commerce on a daily basis.

2. *Mathsya Kiran*: The application was created in 2019 as part of a Masters Research program at ICAR-CIFE (Fig. **4**). It serves as a portal to data on inland fisheries in both text and eye-catching image formats.

3. *Aqua Farmer App*: In 2018, Aqua App created this app (Fig. **5**). The app enables farmers to access the most recent market pricing, information on best practices, news, events, and aquaculture-related videos. Available in Hindi, Telugu, and English.

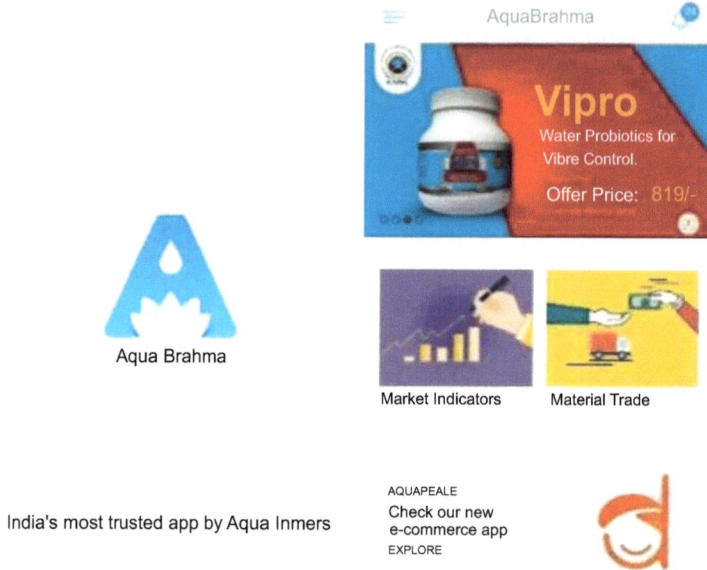

**Fig. (3).** Aqua Brahma App.

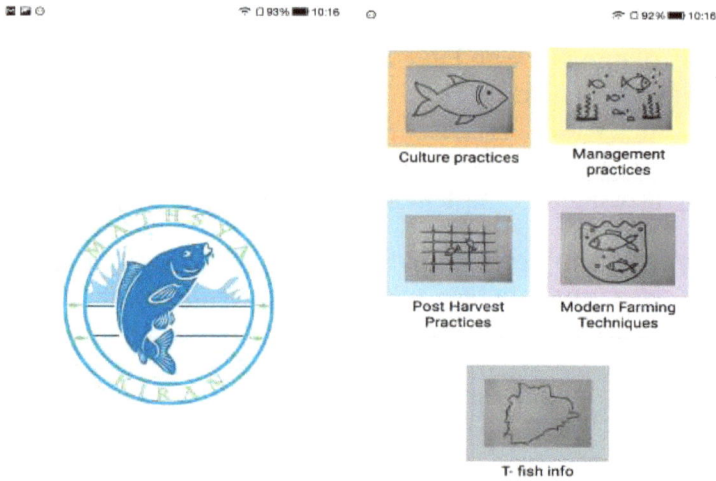

**Fig. (4).** Mathsya Kiran App.

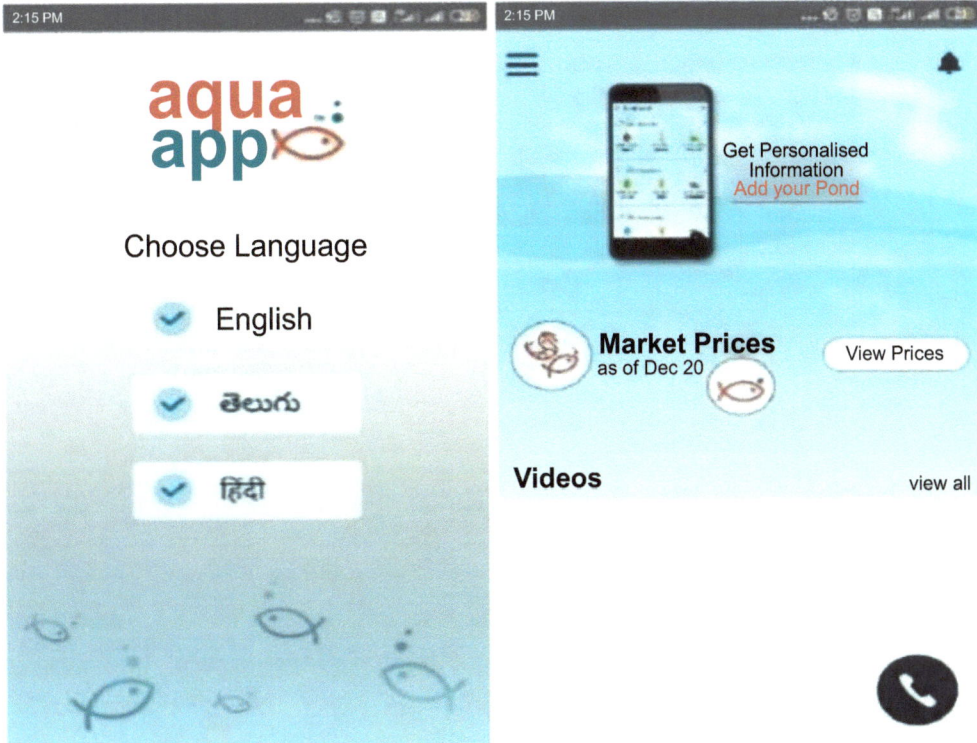

**Fig. (5).** Aqua Farmer App.

4. ***Aquall App***: In 2016, Aquall Foods and Products Pvt. Ltd. created an app. A platform for online shopping that serves as a one-stop shop for agricultural supplies such as seeds, feeds, pesticides, aerators, and tools. Available in both English and Telugu (Fig. **6**).

5. ***Fish Disease Advisory***: ICAR-CIFRI created the app in 2018 (Fig. **7**). The app provides English information about illnesses, their causes, and treatment options.

6. ***Fish Names***: Drapes created the app in 2017 (Fig. **8**). It aids in the offline identification of more than 120 and 72 species of marine and river fish, respectively. The English version of the app is available.

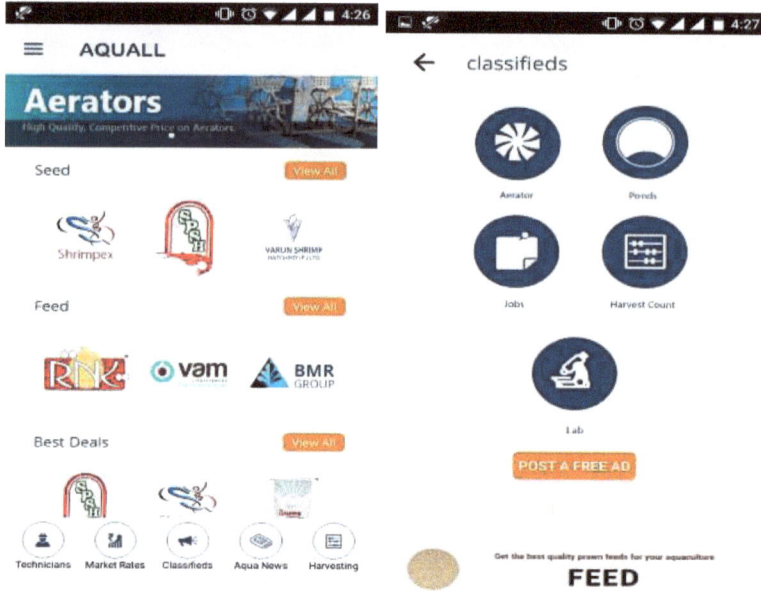

**Fig. (6).** Aquall App.

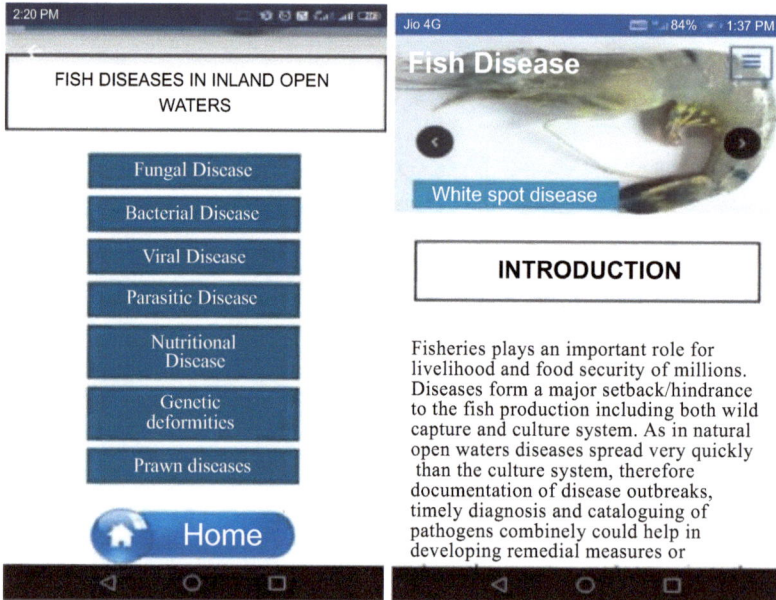

**Fig. (7).** Fish Disease Advisory App.

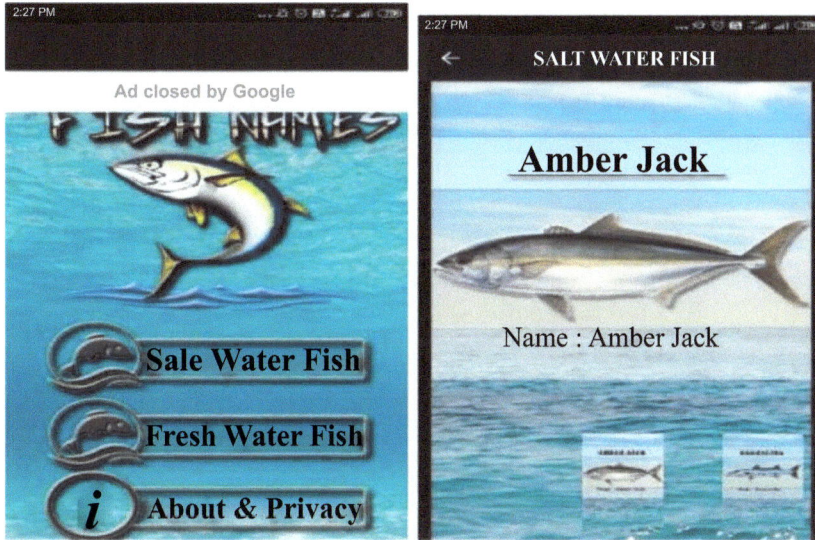

**Fig. (8).** Fish Names App.

7. ***India Aqua***: ICAR-CIFA created the app in 2019. The app serves as an information hub for Indian aquaculture and includes technological modules for various fish species, a database of stakeholders, frequently asked questions, updates, and a discussion forum (Fig. **9**). The app supports Hindi, English, and Odia.

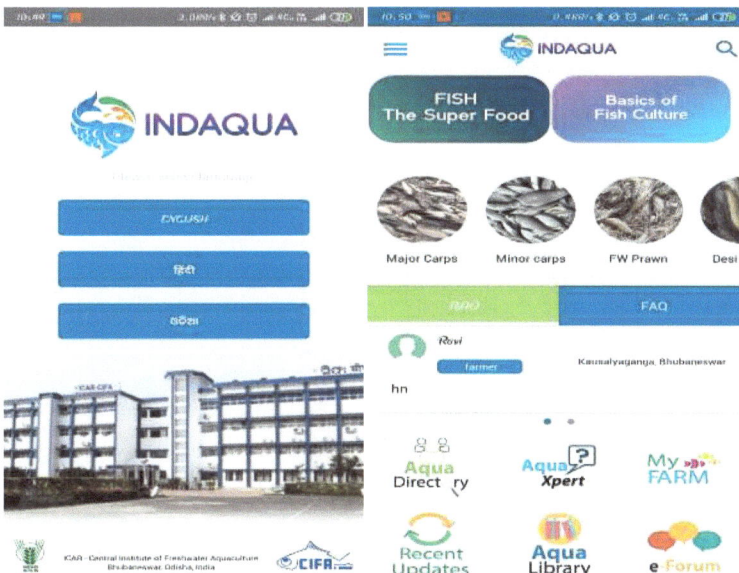

**Fig. (9).** IndAqua App.

8. *Vanamei Shrimp App*: The app was created in 2017 by ICAR-CIBA. Technical details on growing Pacific white shrimp are provided, including optimal management techniques, input calculators, illness diagnostics, risk assessment for shrimp farms, and English FAQs (Fig. **10**).

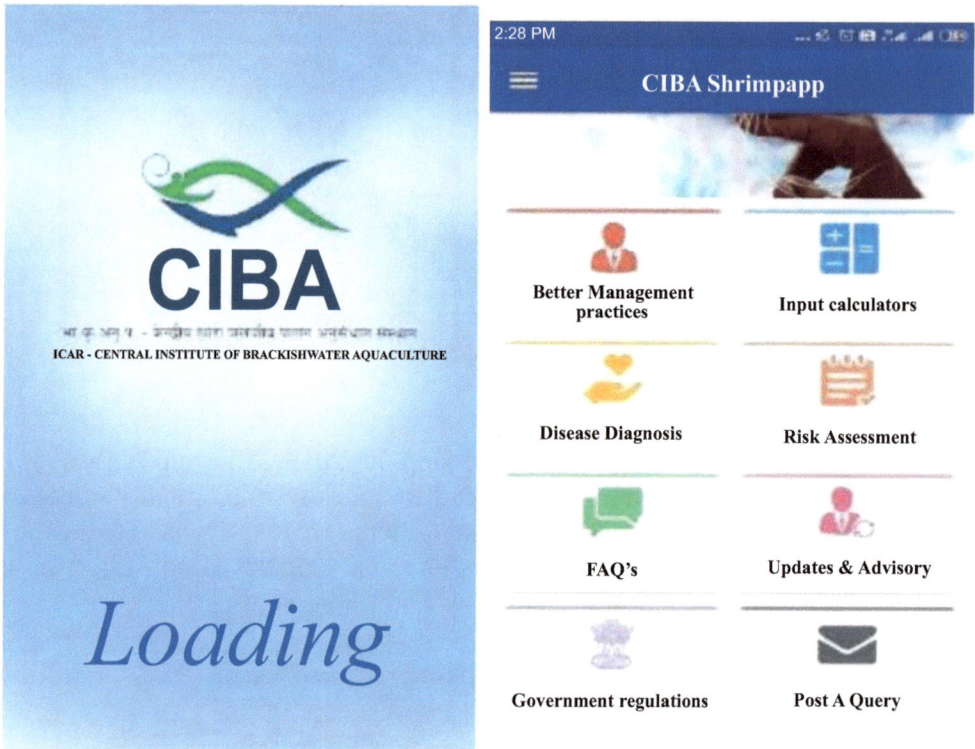

**Fig. (10).**  Vanamei Shrimp App.

9. *Pescare*: It was created in 2017 by budding developers. It is available in English and aids in the detection and treatment of fish and shrimp diseases (Fig. **11**).

10. *mJhinga*: The app was created in 2019 by ICAR-CIFE. It offers details on building additional ponds, raising a productive shrimp crop, and market values (Fig. **12**). To keep track of daily inputs, harvests, and expenses, it has a computerized notebook.

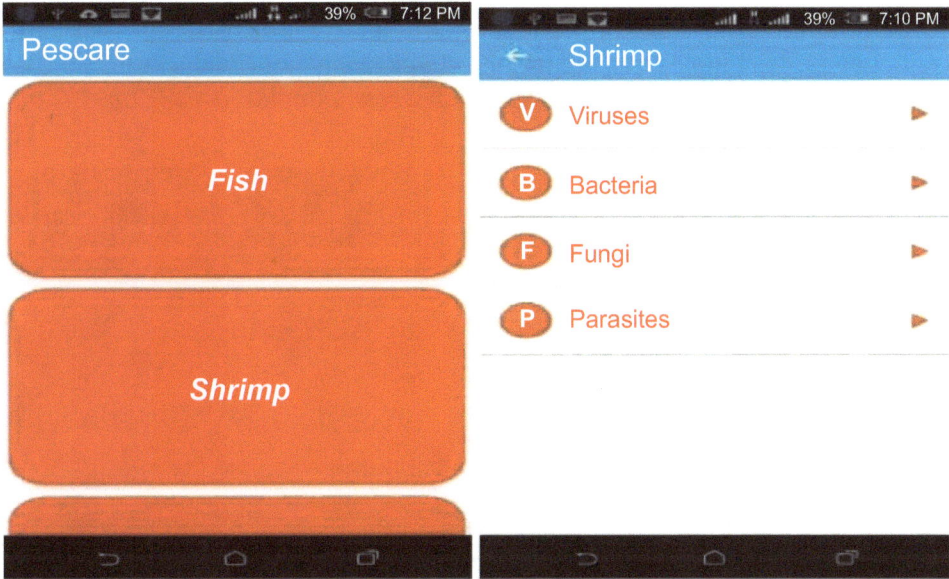

**Fig. (11).** Pescare App.

**Fig. (12).** mJhinga App.

11. **CIFT Lab Test**: Developed in 2019 by ICAR-CIFT, the app aims to provide information on numerous sample testing and analysis of different fish and fish-based goods, fishing gear materials, packaging materials, microbiological parameters, quality parameters of ice and water samples, *etc.* (Fig. **13**).

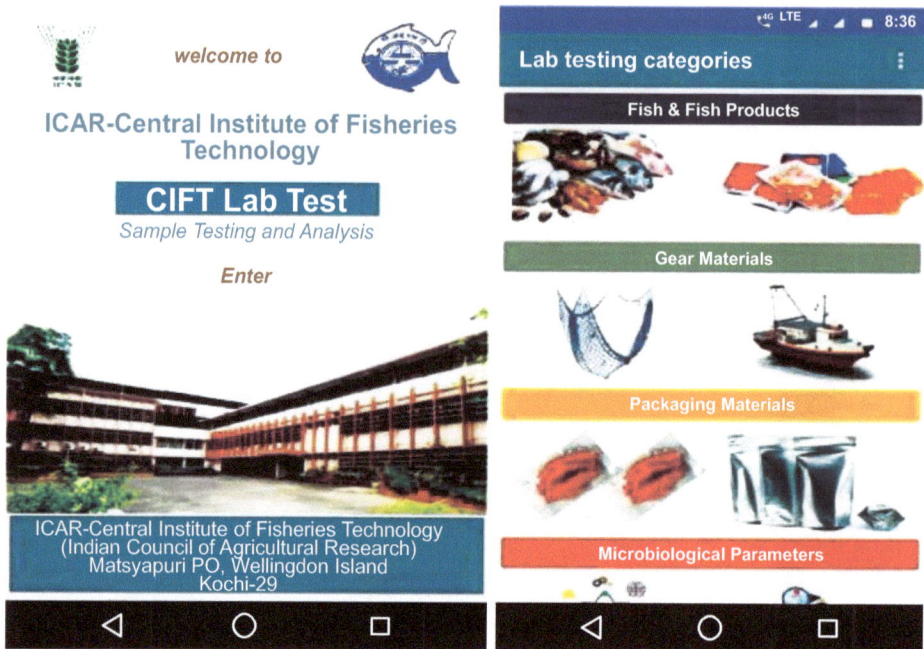

**Fig. (13).** CIFT Lab Test App.

12. **CIFT Training**: In 2019, ICAR-CIFT created the app, which offers a comprehensive collection of information about ICAR-CIFT training programs (Fig. **14**). Students studying fisheries, researchers working in the field, employees of the state extension service and other stakeholders can obtain online information about the many training programs offered by CIFT *via* the app.

## Marine Fisheries-Related Mobile Apps

1. **FFMA (Fisher Friend Mobile App)**: Created in 2016 by the M.S. Swami Nathan Research Foundation. PFZ, Ocean State Forecast, Danger Zone Alerts, Disaster Alerts, and SOS Knowledge Requirements are all addressed by FFMA, a single window solution (Save our Soul) (Fig. **15**). In addition to the Standard English version, the app has been developed in several regional languages of coastal India, including Tamil, Malayalam, Odia, Bangla, Kannada, Marathi, and Gujarati.

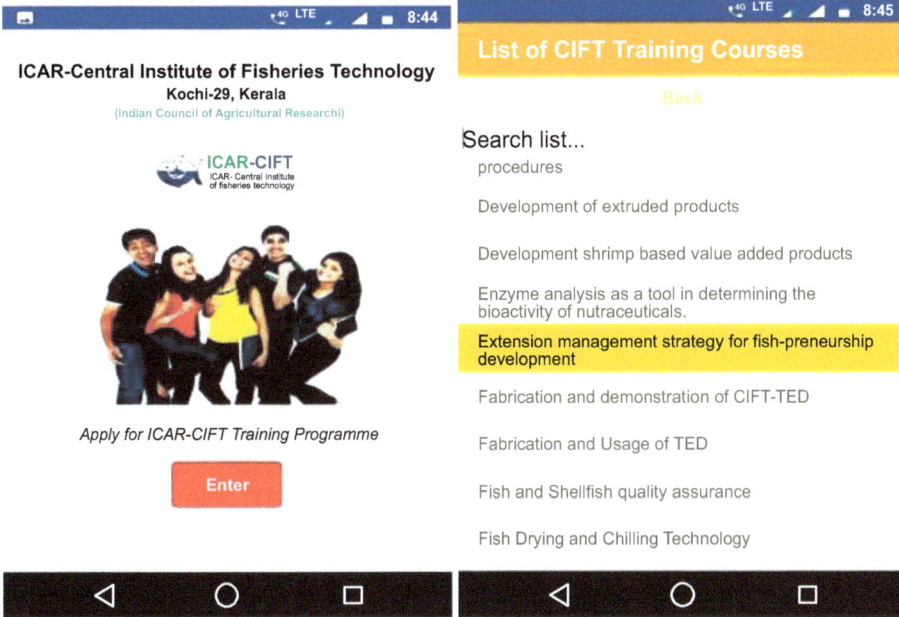

**Fig. (14).** CIFT Training App.

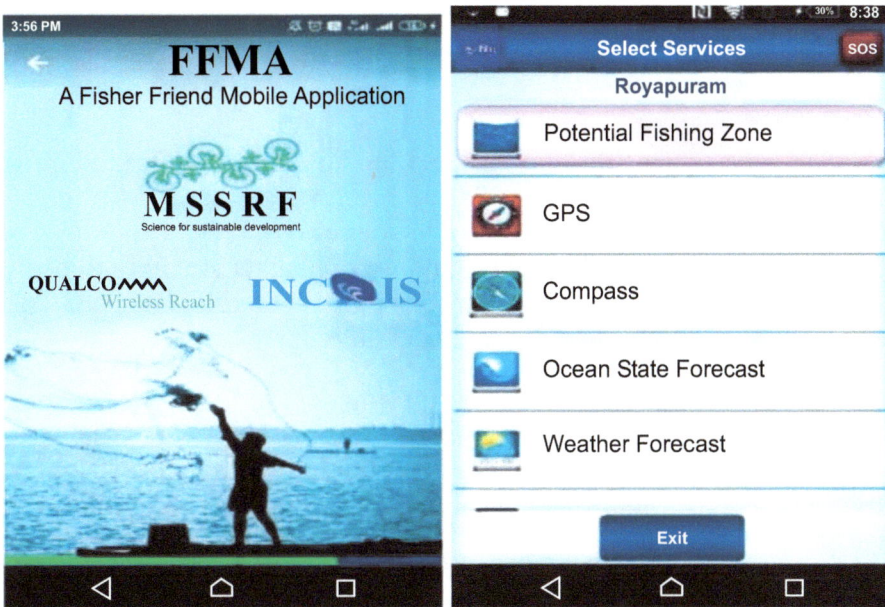

**Fig. (15).** FFMA App.

2. *INCOIS*: In 2016, INCOIS (Indian National Centre for Ocean Information Services) created an app. This software provides English-language access to ocean information services like PFZ and TUNA PFZ for the benefit of the Indian fishing community (Fig. **16**).

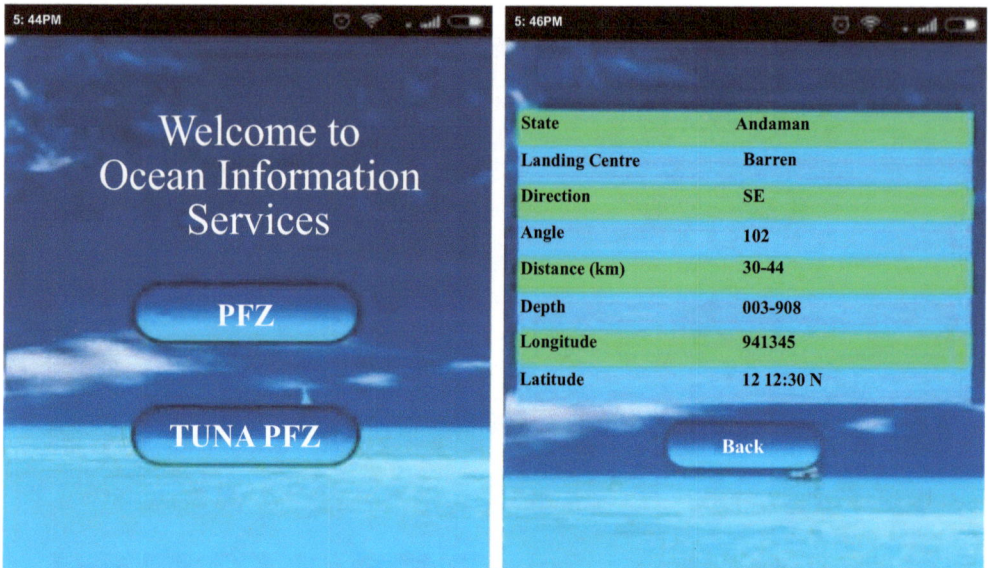

Fig. (16). INCOIS App.

3. *mKrishi@fisheries*: In 2017, the Indian National Center for Ocean Information Services (INCOIS) and the Central Marine Fisheries Research Institute (ICAR) worked together to develop the mKrishi@fisheries app (Fig. **17**). A prospective fishing zone, wind direction and speed, and wave heights are all provided by the app. The Telugu, English, Tamil, Malayalam, Odia, Bangla, Kannada, Marathi, and Gujarati versions of the app are also accessible.

4. *Machli*: In 2019, Reliance Foundation Information Services created an app. At a distance of 50 km from each fish landing location spread out over the whole Indian coastline, the app gives an ocean state forecast for Indian fishermen (Fig. **18**). English, Gujarati, Malayalam, Tamil, Telugu, Kannada, Marathi, and Hindi are the accessible languages.

5. *Odaku*: Odaku Online Services Pvt. Ltd. developed the app. An online service platform for Indian fishermen is provided by the app. In order to support fishermen in their daily activities, it offers tools and technologies (Fig. **19**). It has features like an integrated GPS, addresses the problem of international borders as

well as local borders set by the state level, allows for the sharing of trawling tracks, the buying and selling of secondhand boats, *etc.* English, Gujarati, Malayalam, Tamil, Telugu, Kannada, Marathi, Hindi are the available languages in the app.

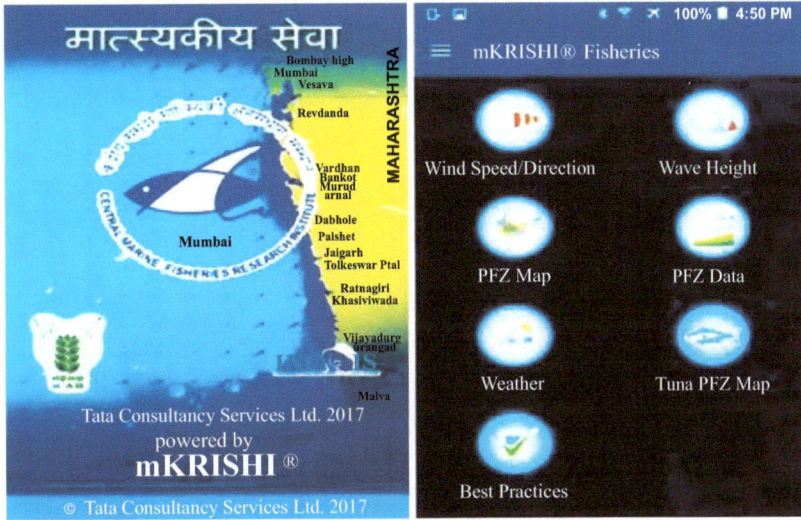

**Fig. (17).** mkrishi@fisheries App.

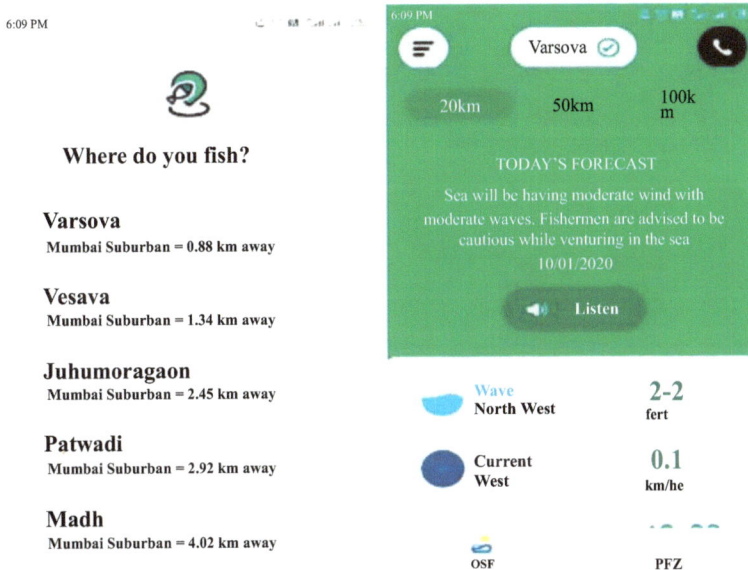

**Fig. (18).** Machil App.

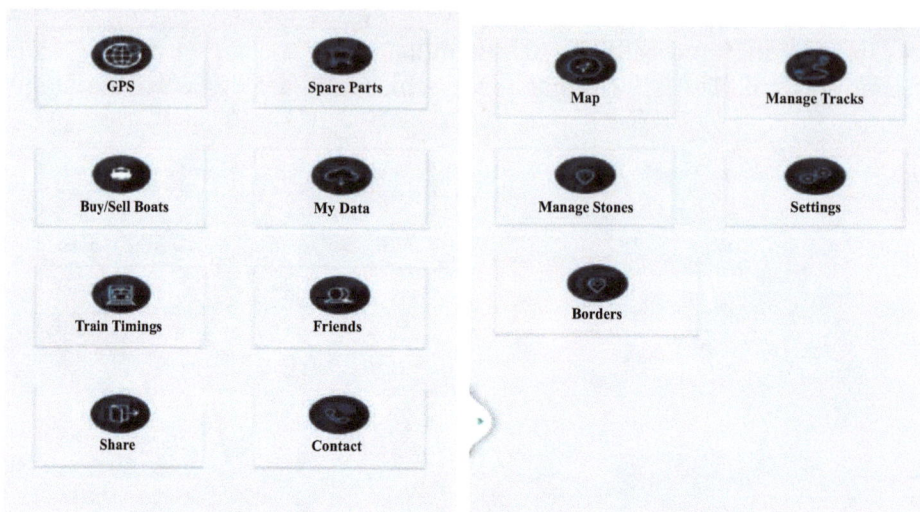

**Fig. (19).** Odaku App.

6. ***PFZ- Advisory***: Designed by Mobile Seva on September 9th, 2015. An app that supports many languages is used to inform fishermen in India's coastal regions about probable fishing zone advisories (Fig. **20**). It is available in English, Gujarati, Malayalam, Tamil, Telugu, Kannada, Marathi, Hindi, and Urdu languages.

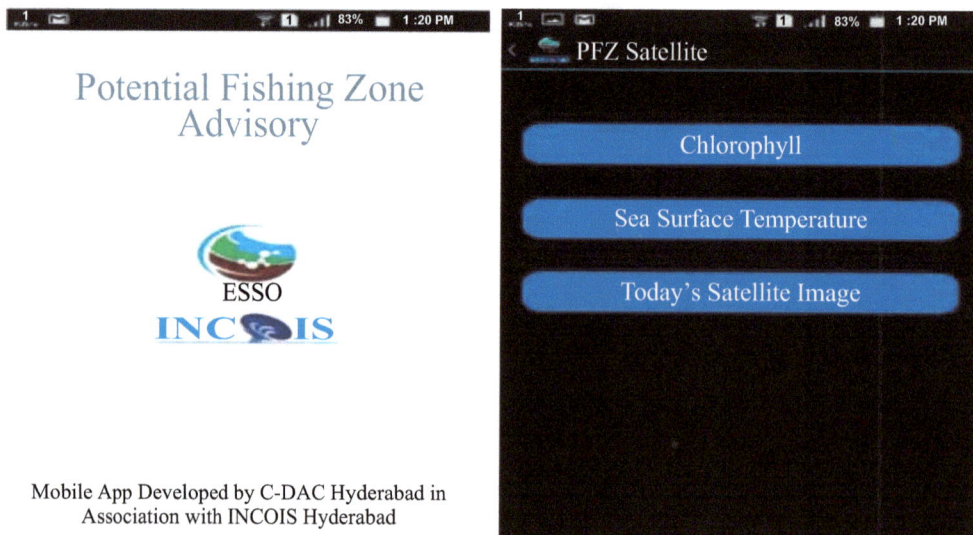

**Fig. (20).** PFZ-Advisory App.

7. ***SARAT (Search and Rescue Aid Tool)***: An app created in 2017 by INCOIS. It is employed in marine research and search and rescue activities (Fig. **21**). Telugu, Kannada, Marathi, Hindi, English, Gujarati, Bangla, Oriya, Malayalam, and Tamil are all available.

**Fig. (21).** SARAT App.

8. ***Sagar Vani***: The app was created in 2017 by INCOIS. It offers information in English regarding PFZ, Ocean state forecast, High wave alerts, and Tsunami early warnings.

9. ***Sagara***: NIC e Gov Mobile Apps created the Sagara app in 2018. The Department of Fisheries and NIC Kerala have launched a digital project to track fishing personnel and boats as they leave for a fishing operation (Fig. **22**). The app is accessible in Tamil, Kannada, Malayalam, and English.

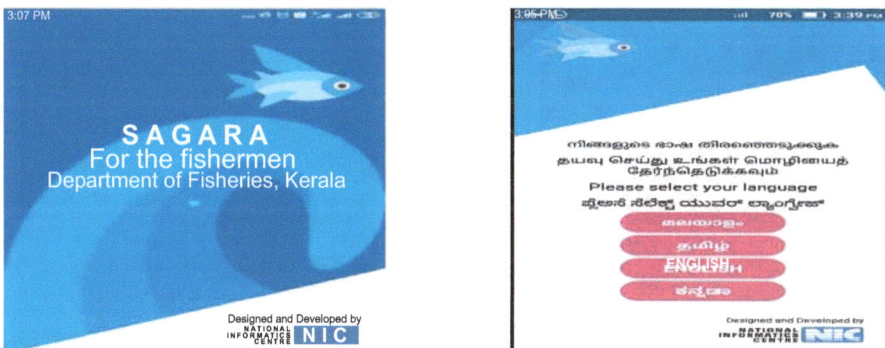

**Fig. (22).** Sagara App.

## Mobile Apps for Marketing

1. *Aqua Pulse*: Pinnacle Soft created the app in 2017. A mobile app for buying shrimp serves as a link between businesses and vendors (Fig. **23**). English language version is available.

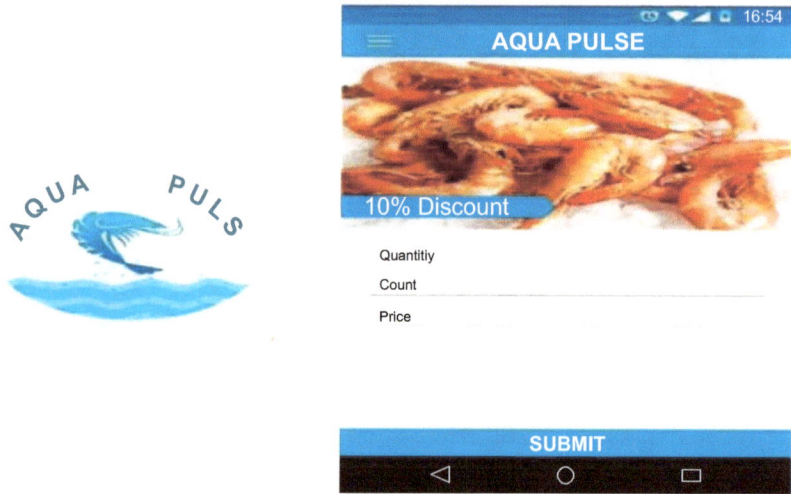

**Fig. (23).** Aquapulse App.

2. *Daily Fish India*: created in 2016 by Baby Marine Enterprise. It is a seafood retailer online (Fig. **24**). In Trivandrum and Ernakulam, daily fish is offered.

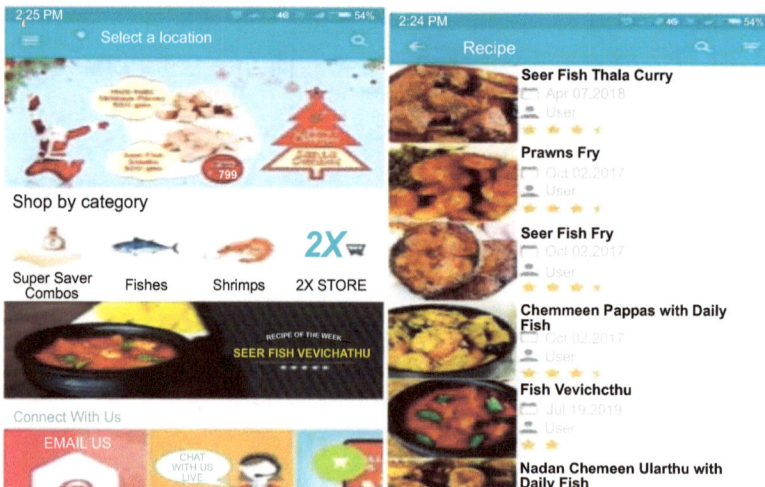

**Fig. (24).** Daily Fish India App.

3. ***Healthy Fish***: In 2016, Healthy Fish created this product. It is an online fish delivery service for farm-raised freshwater fish, dry fish, crabs, shrimp, and lobsters (Fig. **25**).

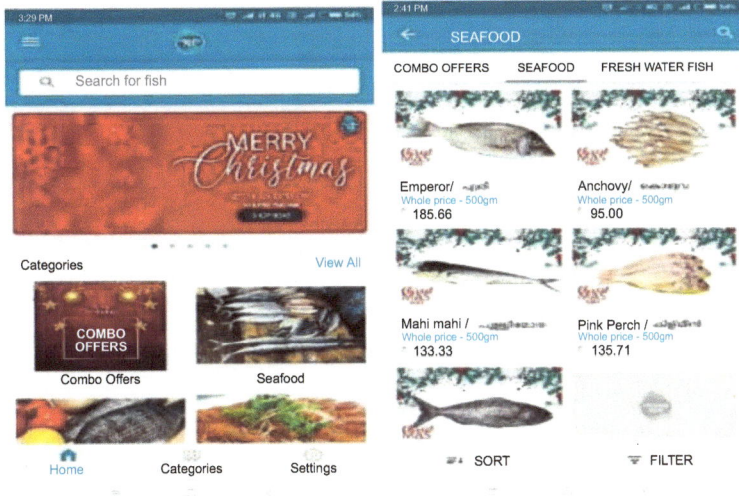

**Fig. (25).** Healthy Fish App.

4. ***Marine fish sales***: In 2019, Zacharia PU and ICAR- CMFRI created an app. Direct sales between fishermen and clients are made possible by the app (Fig. **26**).

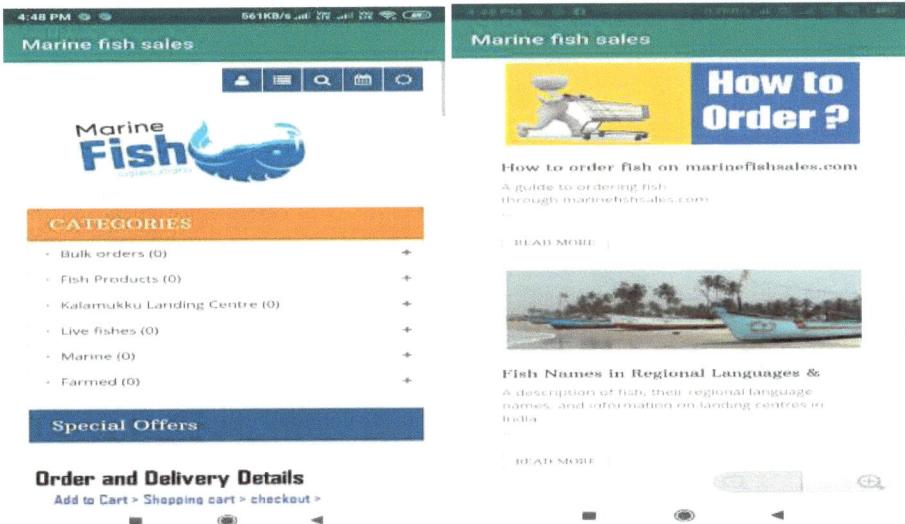

**Fig. (26).** Marine Fish Sales App.

5. *Matha Fresh Fish*: In 2018, Matha Fresh Fish created an app. It is a Keralan online seafood retailer (Fig. 27).

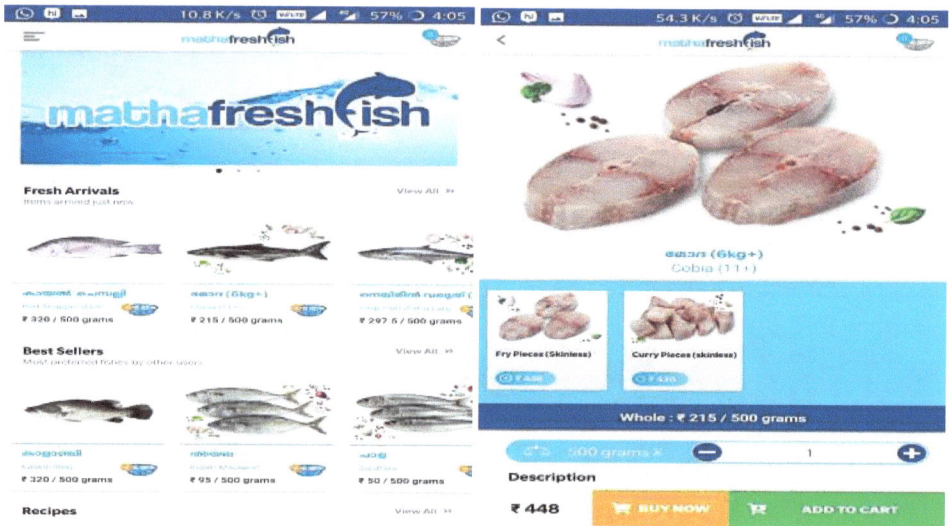

**Fig. (27).** Martha Fresh Fish App.

6. *Nallameen.com*: The app was created in 2017 by Infametech Solutions Pvt. Fresh fish from the daily catch in and around Kochi is delivered using the app (Fig. 28).

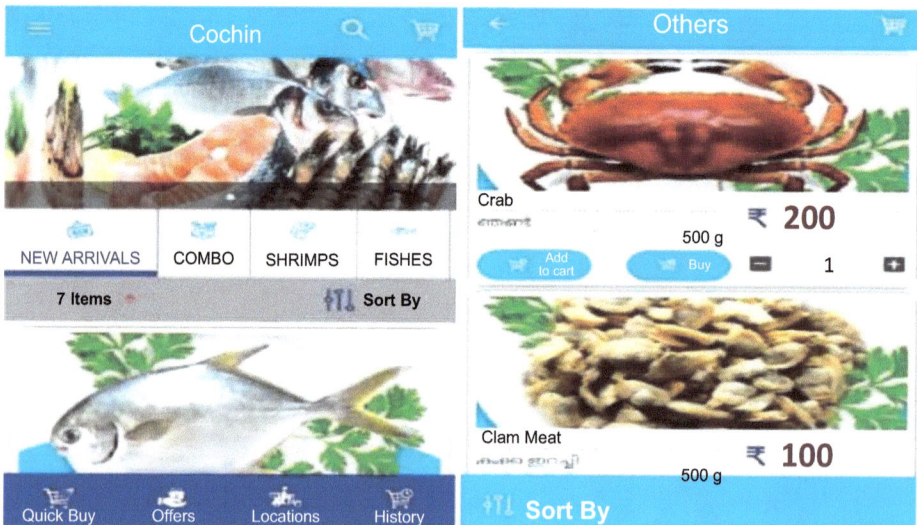

**Fig. (28).** Nallameen.com App.

7. ***Smart fish***: Created by the West Bengal State Department of Fisheries in 2018. It is a fish delivery app available online (Fig. **29**).

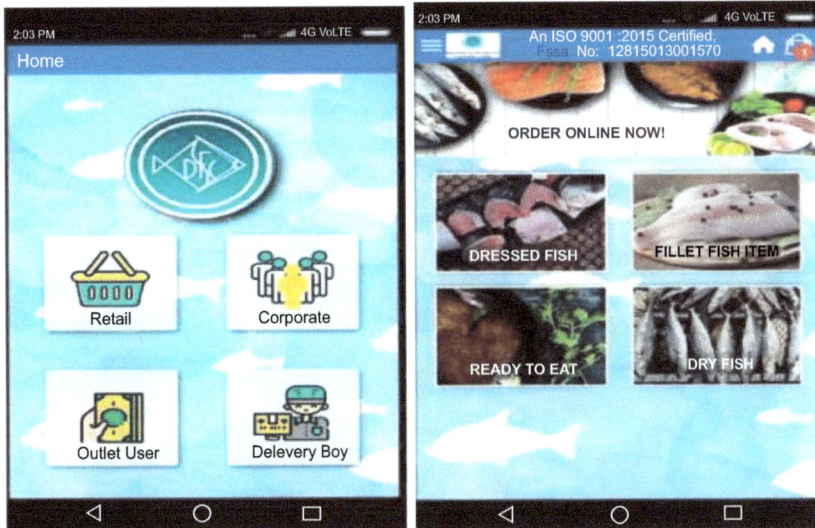

**Fig. (29).** Smart Fish App.

8. ***Fresh Fish Cart***: Astin Soft Pvt. Ltd. created this in 2017. It sends out fresh seafood of every kind (Fig. **30**).

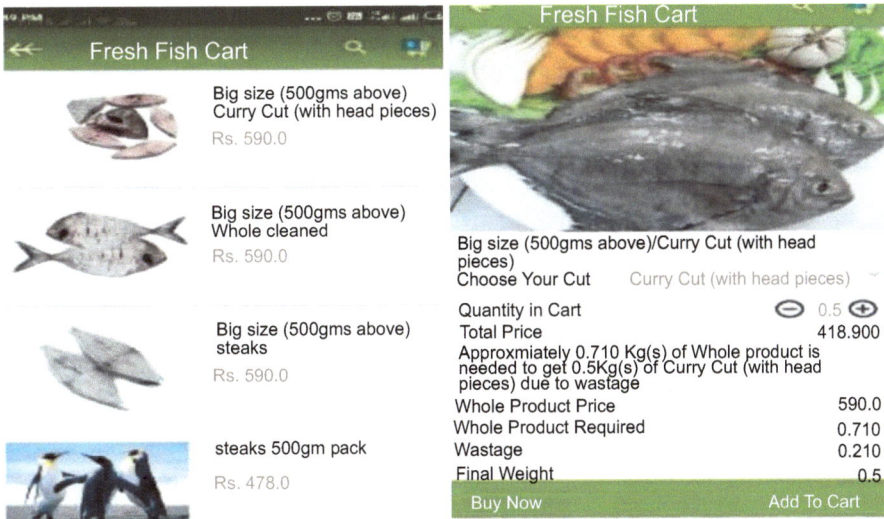

**Fig. (30).** Fresh Fish Cart App.

9. *Aqua Deals*: Mile Deep Works Pvt. Ltd. created the Aqua Deals App, a farmer's market app that sells feed, equipment, health care items, and other items in 9 regional languages (Fig. **31**).

**Fig. (31).** Aqual Deals App.

# FOOD SUPPLY CHAIN MANAGEMENT IN THE AGE OF DIGITALIZATION

A new paradigm that presents opportunities can be seen in the digitalization of operations and supply networks. One significant flexibility-enhancing factor that results from recent developments in industries like information technology, business analytics, and additive manufacturing is digitalization [36, 37]. The digitalization of the supply chain is thus anticipated to be greatly facilitated by IoT. Information technologies created to support modern-day activities are currently proliferating. According to [38], many applications call for a combination of recently developed new technologies, which is spawning creative concepts made possible by technologies like cloud computing, cyber-physical systems, and the Internet of Things (IoT). As a result, in such a setting, effective supply chain management calls for the usage of tools and frameworks [39].

Supply chain management is one of the many sectors where the Internet of Things (IoT) revolution is bringing about a paradigm shift [40, 41]. Devices that are heterogeneously connected will make up the Internet of the future, extending the boundaries of the world with both physical and virtual elements [42]. The IoT is based on developments in sensor technology and widely available broadband communication. Many of the objects in our environment will be connected to the network in some way under the IoT paradigm. According to Hrouga [43], the Internet of Things is built on the integration of sensors/actuators, radio-frequency identification (RFID) tags, and communication technologies. They described how various physical objects and devices that are all around us can be connected to the Internet and used to collaborate and communicate with one another in order to accomplish shared objectives. Companies will spend roughly £250 billion annually on IoT by 2020, with the manufacturing, transportation, and utilities sectors accounting for half of all IoT spending [37]. Manufacturing organizations' conventional business models are significantly impacted by the introduction of the Industrial Internet of Things (IIoT) [44, 45]. Additionally, it is now well-acknowledged how the Internet of Things and new paradigms like Industry 4.0 are related [46, 47].

IoT solutions are better able to address the needs of the supply chain in various industries thanks to advancements in sensor technology. However, managing data and energy consumption for an expanding network of sensors can be difficult. The problems with Internet of Things (IoT) solutions that rely on the usage of sensors for readings and data collecting appear to be best addressed by the Wireless Sensor Network (WSN) Theory [48]. In several applications, including ecosystem monitoring, ocean monitoring, and surveillance, the WSN theory has been used to gather data about physical events [37]. The development of WSNs, according to [49], has made ubiquitous connectivity conceivable and can be seen as one of the main smart sensor technologies that will power the IoT in the future. Furthermore, data collecting is the foundation for further sophisticated applications in IoT environments, according to [39], who also claim that it is the fundamental use of WSN. According to the same experts, IoT produces a lot of data because it is connected to numerous wireless sensor devices. The food supply chain can be significantly impacted by IoT, especially when sensors are used for readings and data collection. The modern food supply chain, according to [50], is highly fragmented and complex; it has a broad geographic and temporal scope, intricate operational procedures, and numerous stakeholders. According to them, complexity has led to a number of problems with operational effectiveness, public food safety, and quality management. The usage of sensors and other IoT-enabled devices plays a significant role in the supply chain for food and perishable items in terms of generating readings of the conditions relevant to the goods being carried.

The fisheries sector may stand to gain the most from the expanding use of IoT solutions involving the use of sensors for readings/data collecting among the numerous types of supply chains accessible in the food business. Increased traceability of fish and seafood products could be utilized to provide more transparency and information for product individualization, as well as to guarantee product quality and compliance with health and safety requirements [41]. Global fish output peaked in 2016 at around 171 million tonnes, according to the Food and Agriculture Organization of the United Nations [51], as reported by [37], with aquaculture accounting for 47% of the total (53% if non-food applications like conversion to fishmeal and fish oil are omitted). Crucial challenges, including waste and freshness, are handled by the fisheries industry. Pelagic fisheries make up the majority of the Mediterranean's fisheries in terms of quantity, according to researchers like [52], as asserted by [37], who studied this issue. Due to their high perishability and the scarcity or inadequacy of ice and freezing facilities, they claimed that waste and spoiling of pelagic fish were significant. IoT-enabled devices can be used to monitor and record any variation in storage and transportation circumstances in this type of situation, which is common in many fisheries across the world.

Weather conditions are constantly affecting the supply chains for food and food production. Changes in weather patterns can have a severe impact on agriculture, as [53] noted while identifying crops affected by a deepening drought and crops growing more quickly as a result of carbon dioxide fertilization. The distribution of sea fisheries in the context of fisheries and seafood is influenced by weather patterns and the water's current state [37]. Since different commercial fish species can be found in littoral zones in waters that are warm, lukewarm, or cold, this is extremely pertinent information. The use of IoT-enabled sensors for readings/data collection of weather conditions for a specific geographic area can provide the opportunity to access substantial and accurate datasets that can be analyzed for the benefit of planning and operations in the fisheries sector in the era of digitalization of food supply chains.

### Technologies that can be Employed in the Context of the IoT for the Supply Chain

IoT stands for Internet of Things, and it was first used by Ashton in 2009 to describe RFID-enabled, individually identifiable, interconnected items [42, 54]. According to [40], "IoT is a network of physical devices that are digitally connected to sense, monitor, and interact within a company and between the company and its supply chain, providing agility, visibility, tracking, and information sharing to facilitate timely planning, management, and coordination of the supply chain activities." The authors' definition focused on the need for digital connectivity of the physical components in the supply chain, as well as the facilitation of data storage, analysis, and sharing; intra- and inter-organizational transactions; and the facilitation of the planning, control, and coordination of supply chain processes. In Fig. (**1**), various IoT-related technologies that can produce enormous volumes of data for the supply chain are schematically represented. These include wireless sensors, smart meters, radio links, RFID tags in trailers, and electronic toll collection (Fig. **32**).

**Fig. (32).** Technologies used in the context of the IoT for the supply chain.

Traffic, emergency services, parking, highways, and other applications can all benefit from data from IoT-enabled devices. IoT produces a lot of data because it is connected to a lot of wireless sensor devices. The adoption of multiple ICT for Supply Chain Management (SCM), including RFID, Enterprise Resource Planning (ERP), and IoT, has grown significantly over the past ten years, according to [43, 55], and as a result, the supply chain has generated a ton of data. The adoption of IT applications has recently been spearheaded by logistics companies. The term "logistics informationization," which was first used by [56], describes the application of contemporary information technology in logistics procedures where a lot of basic data must be entered and processed.

Supply chain and logistics operations have actively adopted IoT solutions in recent years. When tracking containers, for instance, RFID tags are typically used. These tags are connected to the containers, crates, and pallets that are part of the cargo, and they are then tracked at various points along the journey [37]. Numerous freight kinds are tracked and traced using RFID tags. The RFID tag can either be carried by the driver or operator of a haulage vehicle or directly connected to the cargo. In addition to RFID, additional technologies, including wireless sensor networks (WSN), barcodes, intelligent sensing, low-power wireless communications, cloud computing, and others, can also be associated with IoT [37]. One of the most crucial logistics duties, which also include planning, controlling, and storing the flow of products and services, is the delivery role [40].

## CONCLUSION

It is well known that one of the fastest expanding agricultural subsectors is fisheries and aquaculture. Human activities, especially those in the aquaculture industry, are made easier by advances in science, technology, and information. The emergence of device communication *via* IoT and AI is the result of the swift advancement of science and technology. The efficiency of time, resources, and production costs, as well as the reduction of negative externalities from aquaculture activities, are all benefits of digitization in the aquaculture industry. The industry can expand more by utilizing cutting-edge technology like mobile apps, blockchain, AI, and IoT. Utilizing the Internet of Things (IoT) technology, a water quality monitoring system might be able to help solve all the issues. All of these new technologies have a huge potential in the market.

## REFERENCES

[1]     E. Barlian, T. Mursitama, Elidjen, Y.D. Pradipto, and Y. Buana, "The influence of entrepreneurship orientation and IOT capabilities to sustainable competitive advantage of artisanal fisheries in Indonesia: A case study of Artisanal Fishery in Banten Province", *IOP Conf. Ser. Earth Environ. Sci.,* vol. 729, no. 1, p. 012034, 2021.
[http://dx.doi.org/10.1088/1755-1315/729/1/012034]

[2]     N. Ya'acob, N.N.S.N. Dzulkefli, A.L. Yusof, M. Kassim, N.F. Naim, and S.S.M. Aris, "Water quality monitoring system for fisheries using internet of things (IoT)", *IOP Conf. Series Mater. Sci. Eng.,* vol. 1176, no. 1, p. 012016, 2021.
[http://dx.doi.org/10.1088/1757-899X/1176/1/012016]

[3]     K. Dhenuvakonda, and A. Sharma, "Mobile apps and internet of things (IoT): A promising future for Indian fisheries and aquaculture sector", *J. Entomol. Zool. Stud.,* vol. 8, no. 1, pp. 1659-1669, 2020.

[4]     M. Islam, J. Uddin, M.A. Kashem, F. Rabbi, and M. Hasnat, "Design and implementation of an IoT system for predicting aqua fisheries using arduino and KNN", In: *International Conference on Intelligent Human Computer Interaction* Springer: Cham, 2020, pp. 108-118.

[5]     A.T. Tamim, H. Begum, S.A. Shachcho, M.M. Khan, B. Yeboah-Akowuah, M. Masud, and J.F. Al-Amri, "Development of IoT based fish monitoring system for aquaculture", *Intell. Autom. Soft. Comp.,* vol. 32, no. 1, pp. 55-71, 2022.
[http://dx.doi.org/10.32604/iasc.2022.021559]

[6]     A. Fauzan'Adziimaa, and G.R. Arta, "Prototype design of IoT water turbidity sensor based for freshwater fisheries", *2021 International Conference on Advanced Mechatronics, Intelligent Manufacture and Industrial Automation (ICAMIMIA),* pp.113-118, 2021.

[7]     N. Thai-Nghe, T.T. Hung, and N.C. Ngon, "A forecasting model for monitoring water quality in aquaculture and fisheries IoT systems", *2020 Int.Conf. Adv.Comp. Appl. (ACOMP),* pp.165-169, 2020.
[http://dx.doi.org/10.1109/ACOMP50827.2020.00033]

[8]     C.K. Yap, S.H.T. Peng, and C.S. Leow, "Contamination in Pasir Gudang Area, Peninsular Malaysia: What can we learn from Kim Kim River chemical waste contamination?", *J. Humanist. Educ. Dev.,*

vol. 1, no. 2, pp. 82-87, 2019.
[http://dx.doi.org/10.22161/jhed.1.2.4]

[9]     J. Janet, S. Balakrishnan, and S.S. Rani, "IOT based fishery management system", *Int. J. Ocean. Oceanog.,* vol. 973, no. 2667, pp. 147-152, 2019.

[10]    K.A. Chy, A.K.M. Masum, M.E. Hossain, G.R. Alam, S.I. Khan, and M.S. Alam, "A low-cost ideal fish farm using IoT: In the context of Bangladesh aquaculture system", In: *Inventive Communication and Computational Technologies.* Springer: Singapore, 2020, pp. 1273-1283.

[11]    C.S. Goud, S. Das, R. Kumar, C.V. Mahamuni, and S. Khedkar, "Wireless Sensor network (WSN) model for shrimp culture monitoring using open source IoT", *2020 Second International Conference on Inventive Research in Computing Applications (ICIRCA),* pp.764-767, 2020.
[http://dx.doi.org/10.1109/ICIRCA48905.2020.9183178]

[12]    W.N. Probst, "How emerging data technologies can increase trust and transparency in fisheries", *ICES J. Mar. Sci.,* vol. 77, no. 4, pp. 1286-1294, 2020.
[http://dx.doi.org/10.1093/icesjms/fsz036]

[13]    L.A. Dobrzański, and A.D. Dobrzańska-Danikiewicz, "Why are carbon-based materials important in civilization progress and especially in the industry 4.0 stage of the industrial revolution", *Mater. Perform. Charact.,* vol. 8, no. 3, p. 20190145, 2019.
[http://dx.doi.org/10.1520/MPC20190145]

[14]    F.M. Pratiwy, M.D. Cahya, and Y. Andriani, "Digitization of aquaculture: A review", *Int. J. Fish. Aquat. Stud.,* vol. 10, no. 1, pp. 18-22, 2022.
[http://dx.doi.org/10.22271/fish.2022.v10.i1a.2623]

[15]    D.B. Solpico, Y. Nishida, and K. Ishii, "Development of current sensors for digitizing expert knowledge in fish feeding towards sustainable aquaculture", *Proc. Int. Conf. Artif. Life. .Robot. (ICAROB2021),* pp.257-261, 2021.
[http://dx.doi.org/10.5954/ICAROB.2021.OS23-3]

[16]    B. Purba, and L.E. Nainggolan, *Natural Resource Economics.,* J. Simarmata, Ed., We Write Foundation: Medan, 2020.

[17]    G.J. Meaden, and J. Aguilar-Manjarrez, "Advances in geographic information systems and remote sensing for fisheries and aquaculture", *FAO Fish. Aquac. Tech. Pap.,* no. 552, p. I, 2013.

[18]    N. Financia Gusmawati, A. Andayani, and U. Mu'awanah, "Pemanfaatan data penginderaan jauh resolusi tinggi untuk pemetaan tambak di kecamatan Ujung Pangkah, Gresik (Utilization of very high remote sensing imagery for aquaculture ponds mapping in Ujung Pangkah District, Gresi)", *J. Kelaut. Nasion.,* vol. 11, no. 1, pp. 35-51, 2016.
[http://dx.doi.org/10.15578/jkn.v11i1.6065]

[19]    M.A. Sandra, Y. Andriani, K. Haetami, W. Lili, I. Zidni, and M.F. Wiyatna, "Effect of adding fermented restaurant waste meal with different concentration to physical quality of fish pellet", *Int. J. Fish. Aquacul.,* vol. 5, pp. 1-7, 2019.

[20]    Y. Andriani, A. A. Hutapea, I. Zidni, W. Lili, and M. F. Wiyatna, "Literature review on fermentation factors of restaurant organic waste affecting feed quality", *Depik,* vol. 10, no. 3, 2021.

[21]    R.D. Rahayani, and A. Gunawan, "Proposed design of an automatic feeder and aerator systems for shrimps farming", *Int. J. Mat.Mechan. Manufact.,* vol. 6, no. 4, pp. 277-280, 2018.

[22]    S. Suratno, and F. Kurniawan, "Automatic feeder for fish feed controlled by hand phone (Teleautofeeder)", *Prosiding Forum Inovasi Teknologi Akuakultur,* pp. 729-732, 2013.

[23]    C. Zhou, K. Lin, D. Xu, C. Sun, L. Chen, S. Zhang, and Q. Guo, "Computer vision and feeding behavior based intelligent feeding controller for fish in aquaculture", In: *Int. Conf. Comp. Comput. Technol. Agricul.* Springer: Cham, 2017, pp. 98-107.

[24]    C. Noble, K. Mizusawa, K. Suzuki, and M. Tabata, "The effect of differing self-feeding regimes on the

growth, behaviour and fin damage of rainbow trout held in groups", *Aquaculture,* vol. 264, no. 1-4, pp. 214-222, 2007.
[http://dx.doi.org/10.1016/j.aquaculture.2006.12.028]

[25]  T. Napaumpaiporn, N. Chuchird, and W. Taparhudee, "Study on the efficiency of three different feeding techniques in the culture of Pacific white shrimp (Litopenaeus vannamet)", *J. Fish. Environ.,* vol. 37, no. 2, pp. 8-16, 2013.

[26]  C. Ullman, M.A. Rhodes, and D. Allen Davis, "Feed management and the use of automatic feeders in the pond production of Pacific white shrimp Litopenaeus vannamei", *Aquaculture,* vol. 498, pp. 44-49, 2019.
[http://dx.doi.org/10.1016/j.aquaculture.2018.08.040]

[27]  F.D. Von Borstel, J. Suárez, E. de la Rosa, and J. Gutiérrez, "Feeding and water monitoring robot in aquaculture greenhouse. Industrial Robot", *Int. J.,* 2013.

[28]  Z. Hu, R. Li, X. Xia, C. Yu, X. Fan, and Y. Zhao, "A method overview in smart aquaculture", *Environ. Monit. Assess.,* vol. 192, no. 8, p. 493, 2020.
[http://dx.doi.org/10.1007/s10661-020-08409-9] [PMID: 32642861]

[29]  R.A. Bórquez López, L.R. Martinez Cordova, J.C. Gil Nuñez, J.R. Gonzalez Galaviz, J.C. Ibarra Gamez, and R. Casillas Hernandez, "Implementation and evaluation of open source hardware to monitor water quality in precision aquaculture", *Sensors,* vol. 20, no. 21, p. 6112, 2020.
[http://dx.doi.org/10.3390/s20216112] [PMID: 33121079]

[30]  L.K.S. Tolentino, C.P. De Pedro, J.D. Icamina, J.B.E. Navarro, L.J.D. Salvacion, G.C.D. Sobrevilla, A.A. Villanueva, T.M. Amado, M.V.C. Padilla, G.A.M. Madrigal, and L.A.C. Enriquez, "Development of an IoT-based intensive aquaculture monitoring system with automatic water correction", *Int. J. Comp. Dig. Sys.,* vol. 10, no. 1, pp. 1355-1365, 2021.
[http://dx.doi.org/10.12785/ijcds/1001120]

[31]  R. Muninggar, and H. Aulia, "Consumer perceptions of digital marketing of fishery products in special areas of the capital city of Jakarta", *ALBACORE J. Penelit. Perikan. Laut.,* vol. 4, no. 3, pp. 257-269, 2021.
[http://dx.doi.org/10.29244/core.4.3.257-269]

[32]  A. Sharma, and K. Dhenuvakonda, "Virtual fisheries through mobile apps: The way forward", *Aquaculture,* vol. 20, pp. 16-12, 2019.

[33]  A.A. Hingorjo, and B. Memon, "Use of mobile phone in fisheries profession: A perspective of the fishermen of the Indus Delta", *AMCAP J. Med. Commun. Stud.,* vol. 1, no. 2, pp. 101-119, 2021.

[34]  K. Dhenuvakonda, A. Sharma, K.P. Prasad, and R. Sharma, "Socio-economic profile of fish farmers of Telangana and usage of mobile apps", *Asian. J. Agricul. Exten.Econ. Sociol.,* vol. 37, no. 3, pp. 1-9, 2019.

[35]  B.E. Agossou, and T. Toshiro, "IoT & AI based system for fish farming: Case study of Benin", *Proc. Conf. Inform. Technol. Soc. Good.,* pp. 259-264, 2021.
[http://dx.doi.org/10.1145/3462203.3475873]

[36]  D. Ivanov, A. Das, and T.M. Choi, "New flexibility drivers for manufacturing, supply chain and service operations", *Int. J. Prod. Res.,* vol. 56, no. 10, pp. 3359-3368, 2018.
[http://dx.doi.org/10.1080/00207543.2018.1457813]

[37]  A.E. Coronado Mondragon, C.E. Coronado Mondragon, and E.S. Coronado, "Managing the food supply chain in the age of digitalisation: A conceptual approach in the fisheries sector", *Prod. Plann. Contr.,* vol. 32, no. 3, pp. 242-255, 2021.
[http://dx.doi.org/10.1080/09537287.2020.1733123]

[38]  L.D. Xu, E.L. Xu, and L. Li, "Industry 4.0: State of the art and future trends", *Int. J. Prod. Res.,* vol. 56, no. 8, pp. 2941-2962, 2018.
[http://dx.doi.org/10.1080/00207543.2018.1444806]

[39]    P. Morella, M.P. Lambán, J. Royo, and J.C. Sánchez, "Study and analysis of the implementation of 4.0 technologies in the agri-food supply chain: A state of the art", *Agronomy,* vol. 11, no. 12, p. 2526, 2021.
[http://dx.doi.org/10.3390/agronomy11122526]

[40]    M. Ben-Daya, E. Hassini, and Z. Bahroun, "Internet of things and supply chain management: A literature review", *Int. J. Prod. Res.,* vol. 57, no. 15-16, pp. 4719-4742, 2019.
[http://dx.doi.org/10.1080/00207543.2017.1402140]

[41]    L.F. Rahman, L. Alam, M. Marufuzzaman, and U.R. Sumaila, "Traceability of sustainability and safety in fishery supply chain management systems using radio frequency identification technology", *Foods,* vol. 10, no. 10, p. 2265, 2021.
[http://dx.doi.org/10.3390/foods10102265] [PMID: 34681313]

[42]    G.M. Razak, L.C. Hendry, and M. Stevenson, "Supply chain traceability: A review of the benefits and its relationship with supply chain resilience", *Prod. Plann. Contr.,* pp. 1-21, 2021.

[43]    M. Hrouga, A. Sbihi, and M. Chavallard, "The potentials of combining blockchain technology and internet of things for digital reverse supply chain: A case study", *J. Clean. Prod.,* vol. 337, p. 130609, 2022.
[http://dx.doi.org/10.1016/j.jclepro.2022.130609]

[44]    D. Kiel, C. Arnold, and K.I. Voigt, "The influence of the industrial internet of things on business models of established manufacturing companies : A business level perspective", *Technovation,* vol. 68, pp. 4-19, 2017.
[http://dx.doi.org/10.1016/j.technovation.2017.09.003]

[45]    A. Rejeb, S. Simske, K. Rejeb, H. Treiblmaier, and S. Zailani, "Internet of things research in supply chain management and logistics: A bibliometric analysis", *Int. Thing.,* vol. 12, p. 100318, 2020.
[http://dx.doi.org/10.1016/j.iot.2020.100318]

[46]    G. Salvini, G.J. Hofstede, C.N. Verdouw, K. Rijswijk, and L. Klerkx, "Enhancing digital transformation towards virtual supply chains: A simulation game for Dutch floriculture", *Prod. Plann. Contr.,* pp. 1-18, 2020.

[47]    H. Golpîra, S.A.R. Khan, and S. Safaeipour, "A review of logistics internet-of-things: Current trends and scope for future research", *J. Ind. Inf. Integr.,* vol. 22, p. 100194, 2021.
[http://dx.doi.org/10.1016/j.jii.2020.100194]

[48]    H. Fatorachian, and H. Kazemi, "A critical investigation of Industry 4.0 in manufacturing: Theoretical operationalisation framework", *Prod. Plann. Contr.,* vol. 29, no. 8, pp. 633-644, 2018.
[http://dx.doi.org/10.1080/09537287.2018.1424960]

[49]    A.E.C. Mondragon, C.E.C. Mondragon, and E.S. Coronado, "Feasibility of internet of things and agnostic blockchain technology solutions: A case in the fisheries supply chain", *2020 IEEE 7th International Conference on Industrial Engineering and Applications (ICIEA),* Bangkok, Thailand, pp. 504-508, 2020.
[http://dx.doi.org/10.1109/ICIEA49774.2020.9102080]

[50]    A.Z. Abideen, V.P.K. Sundram, J. Pyeman, A.K. Othman, and S. Sorooshian, "Food supply chain transformation through technology and future research directions : A systematic review", *Logistics,* vol. 5, no. 4, p. 83, 2021.
[http://dx.doi.org/10.3390/logistics5040083]

[51]    FAO (2018). The State of World Fisheries and Aquaculture 2018. Food and Agriculture Organization of the United Nations. Accessed from: http://www.fao.org/state-of-fisheries-aquaculture.

[52]    R. Mbarki, S. Sadok, and I. Barkallah, "Quality changes of the Mediterranean horse mackerel (Trachurus mediterraneus) during chilled storage: The effect of low-dose gamma irradiation", *Radiat. Phys. Chem.,* vol. 78, no. 4, pp. 288-292, 2009.
[http://dx.doi.org/10.1016/j.radphyschem.2008.12.004]

[53] Tol, R. S. (2020). The economic impacts of climate change. Review of Environmental Economics and Policy.

[54] K. Ashton, "That 'internet of things' thing", *RFID J.,* vol. 22, no. 7, pp. 97-114, 2009.

[55] D. Arunachalam, N. Kumar, and J.P. Kawalek, "Understanding big data analytics capabilities in supply chain management: Unravelling the issues, challenges and implications for practice", *Transp. Res., Part E Logist. Trans. Rev.,* vol. 114, pp. 416-436, 2018.
[http://dx.doi.org/10.1016/j.tre.2017.04.001]

[56] Z. Mingxiu, F. Chunchang, and Y. Minggen, "The application used RFID in third party logistics", *Phys. Procedia,* vol. 25, pp. 2045-2049, 2012.
[http://dx.doi.org/10.1016/j.phpro.2012.03.348]

CHAPTER 8

# Tea Rhizospheres and Their Functional Role in Tea Gardens

**Rwitabrata Mallick**[1,*]

[1] *Department of Environmental Science, Amity School of Life Science, Amity University, Madhya Pradesh, India*

**Abstract:** Darjeeling tea (*Camellia sinensis*) is famous worldwide for its excellent aroma and taste, and out of that, the best quality tea is produced in the Kurseong hill area. A year-long analysis of soil samples collected from different sites was done by applying the plate-count method in terms of colony-forming units for determining the presence of microbial population within soils cultivating tea on a monthly basis. Coordination in specific microbes might also be responsible for the impact. Fungi, bacteria, and actinomycetes – these three groups of microbes were tested during the process. Results showed that the neo tea plant, rhizosphere and rhizosphere of several other perpetual plants of various ages, flourishing in age-old tea gardens, seemed to expedite the growth of microbes. At present, the tea rhizosphere has been tested thoroughly, specifically in relation to plant-microbe response. Counter to the common outcomes, rhizosphere and soil ratios were found to be regularly less than 1 in samples collected from age-old tea gardens, showing an overall -ve rhizosphere impact. The finding of the 'negative rhizosphere effect' in old tea bushes is a significant and novel nature of the tea rhizosphere. The -ve impact on the rhizosphere of aged tea bushes does not seem to be a regular phenomenon that is related to the aging of plants generally but might be distinctive particular to tea plants. Other important and associated features include colonization of tea, rhizosphere, soil pH, *etc*. Supremacy of a certain population of microbes, an affinity towards a section of general opponents constitutes a good instance of reciprocated selection in the natural environment. These discoveries have unlocked newer paths for extended research in the field of 'rhizosphere microbiology'. The present study is an attempt to evaluate the transforming features coupled with the microbial activity and diversity in the tea rhizosphere and significant implications in the tea industry.

**Keywords:** Environment, Microbiology, Rhizosphere, Soil, Tea.

\* **Corresponding author Rwitabrata Mallick:** Department of Environmental Science, Amity School of Life Science, Amity University, Madhya Pradesh, India; E-mail:rmallick25088@gmail.com

S. Gowrishankar, Hamidah Ibrahim, A. Veena, K. P. Asha Rani & A. H. Srinivasa (Eds.)

## INTRODUCTION

Darjeeling tea is produced mainly in the tea gardens situated in the Kurseong hill area. As a move towards protecting the brand value of this tea across the globe, the Govt. of India has recently issued a notification for Geographical Indication (GI) under the country's Geographic Indicator of Goods Registration and Protection Act [1]. Those found guilty under the act can now be imprisoned for a period of up to three years, along with a fine of Rs. 25 Lakh. This will definitely boost the exports of Darjeeling Tea and thereby enhance the popularity of Kurseong tea gardens and their eco-tourism potential.

The world-famous Darjeeling Tea is produced amongst the tea gardens spread in and around the Kurseong hill area with an altitude of 600 -2000 meters from mean sea level [2]. For the growth of perfect tea, a minimum of 50-60inch rainfall per annum is required [3]. Considering these climatic factors, the Kurseong hill area is the ideal place for tea plantation. The soothing chilly weather, the quality of soil, the quantity of rainfall, and the suitable sloping terrain have provided Darjeeling Tea with unique "Muscatel flavour" and "Exquisite Banquet" [4]. Umpteen positive natural factors help Darjeeling Tea produced in Kurseong, and its unparalleled uniqueness cannot be found anywhere else throughout the world. Thus, it is the most sought and highly valued. 80% of the total produce is exported every year to the affluent Western and Japanese markets [5]. In and around the Kurseong hill area, there are 32 Tea Estates located. Around 1900-hectare area is covered by these tea estates with a production amount of around 4 million kilogram of tea [6].

*Camellia sp.* is a valuable economical plant cultivated broadly in tropical and subtropical Asian countries. But, dramatically, the production, minimization, and degradation of soil for monoculture of tea bushes for a longer period have evoked an enhancing interest among various pedologists and ecologists [7]. Consecutive problems of monoculture were induced by 3 factors, namely - autotoxicity of root exudates, imbalance of soil nutrients, and microbial community shifting [8]. At present, soil degradation of tea gardens is well recorded from the aspects of physico-chemical properties, which include accumulating aluminium, diminishing pH of soil, hydroxybenzene, fluorin, damage of soil structure and leaching of nutrients [9].

In the year of 1904, Lorenz Hiltner, a plant physiologist and German agronomist, for the first time mentioned the term "rhizosphere". It is named after the Greek word "rhiza", which means root [10]. At present, the definition of the rhizosphere is scrutinized to include three zones as per their relative closeness to, and thus inspired from the root. A year-long analysis of edaphic samples gathered from different sites was done by applying the plate-count method in terms of colonyforming units for determining the presence of microbial population within soils cultivating tea on a monthly basis [11].

## METHODOLOGY

### Withering

The withering process starts with plucking the tea leaves. In this process, the leaves become soft and release water. Normally, with the help of hands, fresh leaves are spread in narrow layers on tats (sections of rough fabric) or trays [12]. Leaves are allowed to wither for 18-20 hours. The time varies according to the meteorological conditions like humidity, temperature, moisture content, and size of the leaf. With the machinery's help, withering time can be reduced, but ultimately, it decreases the quality of the end product [13].

### Rolling

The rolling process starts after withering, where the withered leaf is twisted out of shape, and leaf cells are burst [14]. As a result, polyphenols mix up with enzymes. Generally, the process is to roll bunches of leaves with two hands or one hand using the table surface [15]. This method is applied until the leaves are twisted, adequately coated with juices and ultimately broken into pieces. Regarding mechanised rolling is concerned, it has a round-shaped table attached in the centre with a cone and along the surface with battens (slats) [16]. A round-shaped box without a base is taken along with a coerced lid that stands on the table. The table along with the box revolves eccentrically in reverse motion, and the leaves kept in the box are twisted and passed over the cone and battens in a way familiar to the rolling by hand [17]. Bunches of leaves that are rolled are then smashed up and shifted. After this, "fine" tea leaves are moved to the fermentation room, and the coarse leaves which are not used yet, are rolled repeatedly. In nations around the world, rolling is relinquished in support of alteration by different machines [18].

The crushing, tearing, and curling (CTC) mechanism is composed of two jagged rollers made of metal, closely associated and having the capacity to revolve at different motions, which twist, cut and tear the tea leaves [19]. The machine has one horizontal barrel with a feed hopper at one end and a plate with perforations at the other end. Pressurised through the barrel by a screw-type moving shaft attached with vanes at the middle, the leaves are twisted by resistor plates on the inner surface of the barrel and are cut at the end plate. CTC does not produce the larger leafy grades of tea [20].

## Fermentation

This process starts with the breaking down of the leaves during the process of rolling and carries on while the rolled leaf is spread on flat surfaces or perforated trays made of aluminium under certain monitored conditions of humidity, temperature and aeration. It is actually not fermentation at all but a continuous chemical reaction [21].

## Drying

In the drying process, high temperature deactivates the polyphenol enzymes and dries the leaf, which has around 3% moisture content. Drying caramelizes sugars, thereby adding fragrance to the end product, and communicates the black colour linked with fermented tea [22].

Kurseong hill area elongated from the south is endowed with triple heritage, viz., the grand heritage of the Himalayas, the technological heritage of the Darjeeling Himalayan Railway (DHR), and the heritage of aromatic tea [23]. This place is already internationally acclaimed as the land of the four T's – Toy, Train, Tea, Timber (Forest and Wildlife) and Tourism [24].

## Rhizosphere

The rhizosphere is considered the prime active part which stimulates a host of landscape and worldwide scale processes. Relatively good knowledge of these techniques seems quite challenging towards monitoring the living animals of the planet and the planet itself [25]. There is a projection of two times more demand for food globally in the coming 50 years, and to achieve that, efforts are being made to harness the root system of plants and enhance the yield potential of staple food crops [26]. These kinds of initiatives are undertaken in light of the changing climate worldwide and population explosion globally, which will undoubtedly need more agricultural food in relatively less fertile land areas. This type of urgency is already being done in several nations [27]. Achieving a worldwide population explosion and change in climatic conditions with improved knowledge and monitoring of the processes of the rhizosphere would be a major scientific understanding in the coming years [28].

The experiment was done on three groups of bacteria, fungi, and actinomycetes. Surprisingly, the result showed below 1 regarding the ratio between rhizosphere and soil [29]. The samples were collected from tea gardens, which clearly indicates an overall negative effect of the rhizosphere. This negative effect can also be caused by specific microorganisms living together [30]. It can be assumed that the rhizosphere of relatively fresh camellia plants and that of various perennial plan5s having variety in age generally developed in tea estates which are already established, could have accelerated microbial growth [31]. The -ve impact on the rhizosphere of relatively much-aged tea gardens does not seem to be a regular incident which is like the aging of plants in normal conditions but might be one of a kind and specific to tea plants [32].

Several experiments were done using soil samples gathered from different tea gardens in India [30]. The results showed various distinguishing characteristics. Tea gardens from where the samples were collected are situated in the eastern Himalayan region, and are characterized by the amount of rainfall, and even some experience snowfall [33].

## DISCUSSION

### Study Area Under Kurseong subdivision

There are altogether 32 (thirty-two) tea gardens distributed among 14gram panchayats of Kurseong Sub-division [34]. The tea gardens are located under two broad headlines:

**The first section is the Kurseong North Section. This section is comprised of the following tea estates -** Balasan Tea Estate, Ambootia Tea Estate, Dilaram Tea Estate, Eden Vale Tea Estate, Ringtong Tea Estate, Moondakotee Tea Estate, Margaret's Hope Tea Estate, Singell Tea Estate, Oaks Tea Estate and Springside Tea Estate [35].

**The second section is the Kurseong South Section. It is comprised of** Makaibari Tea Estate, Giddhapahar Tea Estate, Castleton Tea Estate, Sipoydhura Tea Estate, Ghoomtie Tea Estate, Rohini Tea Estate, Jogomaya Tea Estate, Longview Tea Estate, Sivitar Tea Estate, Mahaldiram Tea Estate, Jungpana Tea Estate, Mohan Majhua Tea Estate, Monteviot Tea Estate, Narbada Majhua Tea Estate, Nurbong Tea Estate, Mullootar Tea Estate, Selim Hill Tea Estate, Dow Hill Tea Estate and Tindharia Tea Estatate [36].

The initial experiments, which were carried out at Makaibari Tea Experimental Garden and Castleton Tea estate, both in Darjeeling District under the Kurseong Subdivision of West Bengal, gave interesting and thought-provoking outcomes [37]. Investigations were carried out for a period of one year, at a monthly interval in the Makaibari tea estate, where stimulation of rhizosphere microbial populations of relatively new tea plants was primarily expected [38]. On the contrary, already established tea bushes with rhizosphere were seen in the area infested by rhizosphere and microbial population of non-rhizosphere area [39]. Microbial analyses of samples obtained from the soil of Castleton tea estate, where the bushes were of assarnica type and the plantations were more than twenty years old, also showed strong inhibition of rhizosphere microbial communities [40].

Bacteria out of three microbial communities, namely, bacteria, fungi, and actinomycetes, was the maximum subdued group among the aged tea rhizosphere [41]. The stimulation of microbes in the rhizosphere (rhizosphere effect) due to plant roots is a popularly known and normal phenomenon and indicates a "positive" impact on rhizosphere microorganisms [42]. The "negative rhizosphere effect", indicating suppression of microbes around the root zone, has been used for the first time to the best of our knowledge, in the case of established tea estates with rhizosphere [43]. It is an important observation and is the general norm. Similar experiments were also conducted in various tea gardens to justify the status of the rhizosphere effect in established bushes [44].

The soil samples were obtained from various locations (1) Rohini tea estate, (2) Ambootia tea estate, and (3) Goomtee tea estate [45]. The tea bushes were of various age groups. While Rohini and Ambootia tea estates represented well-maintained tea plantations, the Goomtee tea estate was not used for a long time [46]. It has been observed that in most of the cases, the R:S ratios signified the same trend as gathered from Makaibari tea plantations, except in the case of estimation of bacterial population from Rohini tea estate, which showed an inhibitory effect even at 4 years of age [47].

The 'negative rhizosphere effect was more prominent in better tea estates. This was found in two tea gardens, namely, Selim Hill Tea Garden and St. Mary's tea estate [48]. Some recent observations performed at Karbia, a newly created tea estate, +ve rhizosphere impact was recorded for actinomycetes and fungi. Various factors might be responsible for the Selim Hill tea estate in terms of minimizing the population of microbes [49].

Most of the tea plants, for their growth and development, prefer soils which are acidic in nature and range in pH between 4.5 to 5.6. Other than this, the report for pH values of tea rhizosphere and the corresponding non-rhizosphere soil samples collected from different locations also justifies that the pH of soil in tea gardens is lowered by tea plantations in a span of twelve months (Table 1). Normally, the pH of the soil with the presence of a rhizosphere was found to be less than the soil without having the presence of a rhizosphere, and the -ve rhizosphere impact seemed to be more prominent in soils given less pH [50]. Minimizing pH might be a matter of fact that is affecting the reduced population of microbes in the developed tea rhizosphere. In in vitro conditions, it has been reported that tea plants help in lowering the pH [51]. Changing of tea roots to a fluid stage resulted in a lowering of pH from 5.5 to 4.2 [52].

Table 1. Relation between age of tea plants in years and soil ph in various study sites.

| Study Site | Age in Years | Soil pH | |
|---|---|---|---|
| - | - | Rhizosphere Soil | Non-Rhizosphere Soil |
| Rohini | 4 | 5.4 | 5.8 |
| - | 8 | 5.4 | 5.9 |
| - | >100 | 4.8 | 5.4 |
| Ambootia | >30 | 4.6 | 5.6 |
| Goomtee | 4 | 4.7 | 5.2 |
| - | 15-20 | 4.5 | 5.2 |
| Makaibari | 32 | 4.8 | 5.5 |
| - | 44 | 4.2 | 5.2 |
| - | 123 | 4.5 | 5.5 |
| Karbia | >100 | 6.0 | 6.1 |

## CONCLUDING REMARKS

Tea is the main plantation crop in the Kurseong hill area, and the finest qualities of tea are produced here. Tea is one of the tourist attractions of this region. With the enhancement of eco-tourism, tea tourism would also play an important role in increasing revenue generation. Research and developmental activity have been done to estimate in which way the microbes are associated with the tea plantations in various tea estates. Several thousand lakhs of microbes live in close association with the soil and plants in tea gardens. This is almost similar to the microorganisms inside the human body.

Tea tourism, if appropriately organized, is expected to upgrade the livelihood of workers of tea gardens. Through tea tourism, visitors to the tea estates will purchase local handicrafts, and folk medicines and can have the essence of local cuisine, which will further pave the opportunity for more national and international tourists, ultimately benefiting local people. The target is to be able to rebuild the rhizosphere community, which is synthetic in nature and can be utilized to enhance water harvesting efficiency, management of nutrients, and sustainability towards climate change. Eco-huts eco-village concept may further be encouraged in the tea estates throughout the Kurseong subdivision for further betterment of eco-tourism. With the help of metagenomic techniques, several microbial species associated with bulk soil, root endophytes, and rhizosphere soil of tea plantations were analysed having a background. From the different results and data obtained, it has been observed that almost eighty percent of the tea rhizosphere and fifty-four percent of the tea root endophytes were unclassified, and which may be part of the unculturable section.

Spread well over the region are the sprawling lush green tea gardens and 9812 ha. of dense forests, including Wildlife, Sanctuaries, a National Park, and varieties of flora and fauna. The proximity of this region to the North Eastern States and to the foreign countries of Nepal, Bhutan, Bangladesh, and Tibet Region of China has considerably enhanced its strategic and tourism importance.

However, in the absence of adequate infrastructure like proper roads, hotels, wayside facilities, and new exotic tourism products, the potential of tourism in the region cannot be fully exploited. It is perhaps pertinent to mention that some of the private tea gardens have already initiated a form of tea tourism in a very isolated manner at the private bungalows/ houses available in gardens like Makaibari Tea Estate. The idea is gradually catching on and gaining popularity by the day as one of the best modes of eco-tourism coupled with leisure and circuit-based activities. Therefore, if properly marketed, tea tourism can form an attractive package for developing rural tourism both for domestic and international tourists.

DHR's last National Standing Committee Meeting highlighted the possibility of Tea-Train Tourism, the concept of which is to ride the DHR to a tea garden, visit a tea factory, roam in a tea garden, meet the pluckers, participate in tea tasting, interact with tea communities in their villages, stay in tea surroundings,\ and visit Tea Research Association/Museum.

Realising the potential of developing Tea Tourism in the region, Government has already relaxed the lease conditions of tea estate lands to allow tea estate owners to utilise their surplus land for developing tourism-related facilities, including accommodation in their tea estates, with prior approval from the Department of Tourism, Government of West Bengal.

Infact of the inability to predict their metabolic needs, these microbes are unculturable. Depending upon the factors of being biotic or abiotic, specific species of microbes dominate a particular rhizosphere. The -ve rhizosphere impact can be observed through the characteristics of long-lived tea plants and the rhizosphere of established tea bushes. The study areas selected are spread in the eastern Himalayan region. Rainfall during monsoon and snowfall during extreme winters are observed in this region. The chemical composition of root exudates, cultural issues, environmental factors and soil characteristics are very much associated with the tea rhizosphere. This, in turn, influences the overall growth and development of particular populations of microbes which are very much adapted to the tea rhizosphere. Information collected from different sample locations showed that different species of *Bacillus* amongst bacteria dominated the established tea rhizosphere. Rhizosphere shows an increasing selectivity with age, with a lesser number of microbes with an increase in age of tea bushes. A perfect example of mutual selection in nature is being shown by a specific microbial population showing dominance over a community of general antagonists. The present study is an attempt to review the differentiating factors linked to the activity and diversity of microbes among tea rhizospheres and their significant applications in the tea industry.

## REFERENCES

[1]     E. Bååth, S. Olsson, and A. Tunlid, "Growth of bacteria in the rhizoplane and the rhizosphere of rape seedlings", *FEMS Microbiol. Lett.,* vol. 53, no. 6, pp. 355-360, 1988.
[http://dx.doi.org/10.1016/0378-1097(88)90501-0]

[2]     G.D. Bowen, and A.D. Rovira, "The rhizosphere", In: *Tropical soil biology and fertility. A handbook of methods.,* J.M. Anderson, J.S.I. Ingram, Eds., CAB International: Aberystwyth, 1989, pp. 101-112.

[3]     C.P. Chanway, and F.B. Holl, "Ecotypic specificity of spruce emergence-stimulating Pseudomonas putida", *For. Sci.,* vol. 39, pp. 520-527, 1993.

[4]     E.A. Curl, and B. Truelove, *The rhizosphere.* vol. 15. Springer Verlag: Berlin, Heidelberg, New York, 1986.
[http://dx.doi.org/10.1007/978-3-642-70722-3]

[5]     A.H. Fitter, and R.K.M. Hay, *Environmental physiology of plants.* Academic Press: New York, 1981.

[6]     L. Hiltner, "Über neuere Erfahrungen und Probleme auf dem Gebiet der Bodenbakteriologie unter besonderer Berücksichtigung der Gründungung und Brache", *Arb Dtsch Landwirtsch Ges,* vol. 98, pp. 59-78, 1904.

[7]     F.B. Holl, and C.P. Chanway, "Rhizosphere colonization and seedling growth promotion of lodgepole pine by Bacillus polymyxa", *Can. J. Microbiol.,* vol. 38, no. 4, pp. 303-308, 1992.
[http://dx.doi.org/10.1139/m92-050]

[8]     K.C. Ivarson, and H. Katznelson, "Studies on the rhizosphere microflora of yellow birch seedlings", *Plant Soil,* vol. 12, no. 1, pp. 30-40, 1960.
[http://dx.doi.org/10.1007/BF01377758]

[9]     L.F. Johnson, and E.A. Curl, *Methods for research on the ecology of soil-borne plant pathogens.* Burgess: Minneapolis, 1972.

[10]   B.S. Kanwar, *Himachal Pradesh agricultural handbook.,* H.P. Vishva Vidyalays, Ed., Palampur, 1990.

[11]   N. Bag, A. Kumar, S.K. Nandi, A. Pandey, and L.M.S. Palni, "Efficient rooting and biological hardening of tissue culture raised tea (Camellia sinensis (L.)O. Kuntze) plants", *In: Proceedings of 2001 International Conference on O-CHA (tea) Culture & Science,* Shizuoka, Japan, pp. 132-135, 2001.

[12]   R. Mallick, "Prospects of eco-tourism", In: *Aranya Saptaha Issue* Govt. Of West Bengal: Forest Department,, 2008, pp. 18-22.

[13]   R. Mallick, "An approach towards natural resource management in kurseong hill area", In: *Banabithi, Wildlife Issue* Govt. Of West Bengal: Forest Department,, 2008, pp. 18-22.

[14]   R. Mallick, "Eco-tourism in Kurseong Hill Areas. Wildly Ours", In: *Director of Information, Department of Information & Cultural Affairs* Govt. Of West Bengal, 2009, pp. 42-47.

[15]   R. Mallick, and S. Rai, "Tea as a mode of sustainable tourism in kurseong hill area, darjeeling, India", *Inter. J. Pharma. Bio Sci.,* pp. 896-900, 2017.

[16]   H. Katznelson, "The rhizosphere effect of mangels on certain groups of micro-organisms", *Soil Sci.,* vol. 62, no. 5, pp. 343-354, 1946.
[http://dx.doi.org/10.1097/00010694-194611000-00001]

[17]   H. Katznelson, "Nature and importance of the rhizosphere", In: *Ecology of soil-borne plant pathogens.,* K.F. Baker, W.C. Synder, Eds., Univ Calif Press: Berkley, 1965, pp. 187-209.

[18]   S. Krupa, and N. Fries, "Studies on ectomycorrhizae of pine. I. Production of volatile organic compounds", *Can. J. Bot.,* vol. 49, no. 8, pp. 1425-1431, 1971.
[http://dx.doi.org/10.1139/b71-200]

[19]   J.M. Lynch, "Microbial interactions in the rhizosphere", *Tsuchi To Biseibutsu,* vol. 30, pp. 33-41, 1987.

[20]   H.P. Bais, S.W. Park, T.L. Weir, R.M. Callaway, and J.M. Vivanco, "How plants communicate using the underground information superhighway", *Trends Plant Sci.,* vol. 9, no. 1, pp. 26-32, 2004.
[http://dx.doi.org/10.1016/j.tplants.2003.11.008] [PMID: 14729216]

[21]   P.N. Benfey, M. Bennett, and J. Schiefelbein, "Getting to the root of plant biology: Impact of the Arabidopsis genome sequence on root research", *Plant J.,* vol. 61, no. 6, pp. 992-1000, 2010.
[http://dx.doi.org/10.1111/j.1365-313X.2010.04129.x] [PMID: 20409273]

[22]   A.G. Bengough, and B.M. McKenzie, "Sloughing of root cap cells decreases the frictional resistance to maize ( Zea mays L.) root growth", *J. Exp. Bot.,* vol. 48, no. 4, pp. 885-893, 1997.
[http://dx.doi.org/10.1093/jxb/48.4.885]

[23]   G. Berg, and K. Smalla, "Plant species and soil type cooperatively shape the structure and function of microbial communities in the rhizosphere", *FEMS Microbiol. Ecol.,* vol. 68, no. 1, pp. 1-13, 2009.
[http://dx.doi.org/10.1111/j.1574-6941.2009.00654.x] [PMID: 19243436]

[24]   B.B. Buchanan, W. Gruissem, and R.L. Jones, *Biochemistry and Molecular Biology of Plants.* Courier Companies Inc.: U.S., 2001.

[25]   F. E. Clark, "Soil micro-organisms and plant roots", *Advances in Agronomy,* vol. 1, pp. 241-288, 1949.

[26]   T. Danhorn, and C. Fuqua, "Biofilm formation by plant-associated bacteria", *Annu. Rev. Microbiol.,* vol. 61, no. 1, pp. 401-422, 2007.
[http://dx.doi.org/10.1146/annurev.micro.61.080706.093316] [PMID: 17506679]

[27]   JR Dinneny, TA Long, JY Wang, JW Jung, D Mace, S Pointer, C Barron, SM Brady, J Schiefelbein, and PN Benfey, "Cell identity mediates the response of arabidopsis roots to abiotic stress", *Science,* vol. 320, no. 5857, pp. 942-945, 2008.
[http://dx.doi.org/10.1126/science.1153795]

[28]   E.S.M. El-Morsy, "Microfungi from the ectorhizosphere-rhizoplane zone of different halophytic plants from the Red Sea Coast of Egypt", *Mycologia,* vol. 91, no. 2, pp. 228-236, 1999.
[http://dx.doi.org/10.1080/00275514.1999.12061012]

[29]   E.F. Estermann, and A.D. McLaren, "Contribution of rhizoplane organisms to the total capacity of plants to utilize organic nutrients", *Plant Soil,* vol. 15, no. 3, pp. 243-260, 1961.
[http://dx.doi.org/10.1007/BF01400458]

[30]   A. Genre, and P. Bonfante, "Check-in procedures for plant cell entry by biotrophic microbes", *Mol. Plant Microbe Interact.,* vol. 20, no. 9, pp. 1023-1030, 2007.
[http://dx.doi.org/10.1094/MPMI-20-9-1023] [PMID: 17849704]

[31]   V. Gewin, "Food: An underground revolution", *Nature,* vol. 466, no. 7306, pp. 552-553, 2010.
[http://dx.doi.org/10.1038/466552a] [PMID: 20671689]

[32]   G.E.D. Oldroyd, and J.A. Downie, "Coordinating nodule morphogenesis with rhizobial infection in

legumes", *Annu. Rev. Plant Biol.,* vol. 59, no. 1, pp. 519-546, 2008.
[http://dx.doi.org/10.1146/annurev.arplant.59.032607.092839] [PMID: 18444906]

[33] J. Handelsman, and E.V. Stabb, "Biocontrol of soilborne plant pathogens", *Plant Cell,* vol. 8, no. 10, pp. 1855-1869, 1996.
[http://dx.doi.org/10.2307/3870235] [PMID: 12239367]

[34] M.J. Harrison, "Signaling in the arbuscular mycorrhizal symbiosis", *Annu. Rev. Microbiol.,* vol. 59, no. 1, pp. 19-42, 2005.
[http://dx.doi.org/10.1146/annurev.micro.58.030603.123749] [PMID: 16153162]

[35] A. Hartmann, M. Rothballer, and M. Schmid, "Lorenz Hiltner, a pioneer in rhizosphere microbial ecology and soil bacteriology research", *Plant Soil,* vol. 312, no. 1-2, pp. 7-14, 2008.
[http://dx.doi.org/10.1007/s11104-007-9514-z]

[36] M.C. Hawes, L.A. Brigham, F. Wen, H.H. Woo, and Y. Zhu, "Function of root border cells in plant health: pioneersin the rhizosphere", *Annu. Rev. Phytopathol.,* vol. 36, no. 1, pp. 311-327, 1998.
[http://dx.doi.org/10.1146/annurev.phyto.36.1.311] [PMID: 15012503]

[37] M.C. Hawes, U. Gunawardena, S. Miyasaka, and X. Zhao, "The role of root border cells in plant defense", *Trends Plant Sci.,* vol. 5, no. 3, pp. 128-133, 2000.
[http://dx.doi.org/10.1016/S1360-1385(00)01556-9] [PMID: 10707079]

[38] H Hellriegel, and H Wilfarth, "Untersuchungen uber die Stickstoffnahrung der Gramineon und Leguminosen. Beilageheft zu der Ztschr", *Ver. Ru¨benzucker-Industrie Deutschen Reichs,* 1888.

[39] L. Hiltner, "Ueber neuere Erfahrungen und Probleme auf dem Gebiete der Bodenbakteriologie und unter besonderer BerUcksichtigung der Grundungung und Brache", *Arb. Deut. Landw. Gesell,* vol. 98, pp. 59-78, 1904.

[40] A.M. Hirsch, M.R. Lum, and J.A. Downie, "What makes the rhizobia-legume symbiosis so special?", *Plant Physiol.,* vol. 127, no. 4, pp. 1484-1492, 2001.
[http://dx.doi.org/10.1104/pp.010866] [PMID: 11743092]

[41] M.D. Ho, J.C. Rosas, K.M. Brown, and J.P. Lynch, "Root architectural tradeoffs for water and phosphorus acquisition", *Funct. Plant Biol.,* vol. 32, no. 8, pp. 737-748, 2005.
[http://dx.doi.org/10.1071/FP05043] [PMID: 32689171]

[42] A. Hodge, "The plastic plant: Root responses to heterogeneous supplies of nutrients", *New Phytol.,* vol. 162, no. 1, pp. 9-24, 2004.
[http://dx.doi.org/10.1111/j.1469-8137.2004.01015.x]

[43] K.M. Jones, H. Kobayashi, B.W. Davies, M.E. Taga, and G.C. Walker, "How rhizobial symbionts invade plants: The Sinorhizobium–Medicago model", *Nat. Rev. Microbiol.,* vol. 5, no. 8, pp. 619-633, 2007.
[http://dx.doi.org/10.1038/nrmicro1705] [PMID: 17632573]

[44] D.L. Jones, C. Nguyen, and R.D. Finlay, "Carbon flow in the rhizosphere: Carbon trading at the soil–root interface", *Plant Soil,* vol. 321, no. 1-2, pp. 5-33, 2009.
[http://dx.doi.org/10.1007/s11104-009-9925-0]

[45] J.W. Kloepper, and M.N. Schroth, *Conf. plant pathogenic bacteria,* vol. 2, pp. 879-882, 1978.

[46] J.W. Kloepper, B. Schippers, and P.A.H.M. Bakker, "Proposed elimination of the term endorhizosphere", *Phytopathology,* vol. 82, pp. 726-727, 1992.

[47] J.W. Kloepper, S. Tuzun, and J.A. Kuć, "Proposed definitions related to induced disease resistance", *Biocontrol Sci. Technol.,* vol. 2, no. 4, pp. 349-351, 1992.
[http://dx.doi.org/10.1080/09583159209355251]

[48] E.M. Knee, F.C. Gong, M. Gao, M. Teplitski, A.R. Jones, A. Foxworthy, A.J. Mort, and W.D. Bauer, "Root mucilage from pea and its utilization by rhizosphere bacteria as a sole carbon source", *Mol. Plant Microbe Interact.,* vol. 14, no. 6, pp. 775-784, 2001.

[http://dx.doi.org/10.1094/MPMI.2001.14.6.775] [PMID: 11386373]

[49]   S. Kosuta, M. Chabaud, G. Lougnon, C. Gough, J. Dénarié, D.G. Barker, and G. Bécard, "A diffusible factor from arbuscular mycorrhizal fungi induces symbiosis-specific MtENOD11 expression in roots of Medicago truncatula", *Plant Physiol.,* vol. 131, no. 3, pp. 952-962, 2003.
[http://dx.doi.org/10.1104/pp.011882] [PMID: 12644648]

[50]   J. Leong, and J.B. Neilands, "Mechanisms of siderophore iron transport in enteric bacteria", *J. Bacteriol.,* vol. 126, no. 2, pp. 823-830, 1976.
[http://dx.doi.org/10.1128/jb.126.2.823-830.1976] [PMID: 131124]

[51]   S. Konishi, N. Tsuji, and T. Kuboi, "Stimulatory effect of aluminium on the growth of cultured roots of tea", *Proceedings of the International Symposium on Tea Science,* Shizuoka, Japan,pp. 742-746, 1991.

[52]   A. Pandey, and M.S.P Lok, "Characteristic features, microbial diversity and applications; topical review", *Int. J. Tea Sci.,* vol. 1, no. 4, pp. 10-24, 2003.

**CHAPTER 9**

# Applications of Smart Farming Sensors: A Way Forward

**Prasenjit Pal**[1,*] and **Sandeep Poddar**[2]

[1] *Department of Fisheries Extension, Economics and Statistics, College of Fisheries, CAU (I), Lembucherra, Tripura, India*

[2] *Lincoln University College, Petaling Jaya, Selangor Darul Ehsan, Malaysia*

**Abstract:** The introduction of sensing-based technology has transformed the agriculture sector in many ways. This chapter explores the potential of sensing-based technology, including big data and artificial intelligence, in agriculture to lower production costs and increase yield efficiencies. The application of various sensors is explained in various sectors of agriculture, like crop farming, animal farming, and fish farming. This technology has the potential to automate farming and has the ability to shift to precise cultivation for higher crop yields and better quality while using the minimum resources. This chapter also elaborates on the different types of sensors used in agriculture, their benefits, and related issues for their various applications. These technologies have some real issues in the application, which need to be sorted out, and more efforts should be made to make the product more cost effective, relevant, and customized for the use of farmers.

**Keywords:** Sensing based technology, Artificial intelligence, Big data and automation.

## INTRODUCTION

The development of the agricultural sector is one of the major ways to minimize extreme poverty, enhance economic prosperity, and feed a projected 9.7 billion people by 2050 [1]. This sector has played a crucial role in the economic prosperity of the developed nations, from employment generation to impacting the National Income. Growth in this sector impacted a significant increase in the per-capita income of the rural community. Thus, greater attention on the agricultural sector is the need of the hour. For countries like India, the agricultural sector accounts for 18% of GDP and provides employment opportunities [2 - 5] for

---

[*] **Corresponding author Prasenjit Pal:** Department of Fisheries Extension, Economics and Statistics, College of Fisheries, CAU (I), Lembucherra, Tripura, India; E-mail:prasenjit3agstat@gmail.com

**S. Gowrishankar, Hamidah Ibrahim, A. Veena, K. P. Asha Rani & A. H. Srinivasa (Eds.)**

nearly 50% of the country's workforce. Development in the agricultural sector in India will confirm rural development, which in turn will lead towards rural transformation [6, 7]. Agriculture is recently experiencing a catalytic transformation due to the advent of sensor-based technology, big data, robotics and artificial intelligence in the sector to enhance the efficacy of farming operations, thereby enhancing farm productivity many times. The introduction of sensing devices for precision agriculture is important to boost crop yields and minimize crop loss. These technologies enrich the data science and analytics needed to perform the different operations of agriculture and eventually optimize its overall efficacy. The growing demand for food all over the world, coupled with the introduction of sensing-based technology [8], has shifted the agriculture sector in many ways, making the modern farm more centralized, large scale and efficient. This chapter explores the potential of various sensors, big data and artificial intelligence to lower production costs, increase efficiencies, enhance animal welfare and grow more animals per hectare. The sector gained impetus from AI based technology to maximize its yield, including solutions for proper soil treatment, disease and pest management, big data requirements, and reducing the knowledge gap between farmers and technology. The advantages of AI in agriculture are its flexibility, high performance, accuracy, and cost-effectiveness. The resistance of farmers towards applying AI is caused by a lack of proper understanding of the practical application of the sensor tools, and sometimes the cost is very high. This is the reality of implementing artificial intelligence in agriculture [9]. Although AI can be useful, technology providers need to play a proactive role to help farmers implement it in the right way. Now the industry, with the help of AI based technologies, is able to grow high yielding crops, control crop diseases and pests in the field, constantly maintain soil and other growing conditions and improve a wide range of activities, from farming to marketing products.

AI based applications have increased the production of crops multiple times and improved real-time monitoring using automation, harvesting, processing, and marketing [10]. Automated systems have contributed largely to the agro-industrial sector. These systems with advanced AI based algorithms are able to detect various weeds, maintain crop quality and use many other techniques [11]. This chapter describes those sensor-based technologies to augment productivity by reducing crop loss [12], describes the importance of various automated soil sensing techniques [13] and explains how temperature and moisture sensors can work together for vehicle predictions.

Sensor based applications have transformed the agricultural sector in many ways [14]. As this sector is the least automated, there is ample opportunity for the commercialization of newly introduced AI based technologies. Artificial

Intelligence (AI) has contributed immensely to changing the environment around us [15 - 18] and providing an alternative solution for harvesting farm crops. There have been various technologies that are embedded and have now become a part of farming; however, IoT sensors used in agriculture are a new type of technology that is being used worldwide now. The use of sensors to acquire environmental and system metrics and offer them to farmers is beneficial for information-based decisions. The smart farming sensors enhance overall output, be it farm animals or crop farming. The real purpose behind the use of smart farming sensors is to provide the growth and enhancement of crop yield by lowering waste and optimizing the use of human labor.

## SMART FARMING: AN EMERGING CONCEPT

Smart farming is a new concept in modern farming that includes the application of sensor-based technology, algorithms of artificial intelligence, automation, and predictive analysis to enhance the output (yield) of the farming system without affecting the environment. Recent automation in farming has replaced human labor. Smart farming is based on recent technologies such as automated machines, sensors, actuators, drones, and security cameras to constantly monitor and operate farming and farm animals. The objective of smart farming is to augment the quality and quantity of farm output (yield) by minimizing cost and energy usage. IoT is capable of modernizing the agricultural field to manage the farming operation in an easier way, process waste generation and provide a significant increase in productivity. A smart farm uses various technologies to help the farmers in different ways, like sensors for soil, water, moisture, and humidity control, devices to detect and control plant and animal disease, and managing and tracking the exact locations of pests/insects using GPS tools [19]. Different researchers are trying to minimize the cost of farming by introducing low-cost devices/sensors (Fig. **1**). Some of the applications of smart farming in the agricultural domain are as follows:

√ Sensors to constantly monitor and track the status of farm crops, animals including insects.

√ Drones for monitoring the livestock, application of fertilizers, insecticides *etc.*

√ Automated/ sensor-based irrigation system based on the needs of the crop.

√ Newly introduced machines for performing farm operation easier than before.

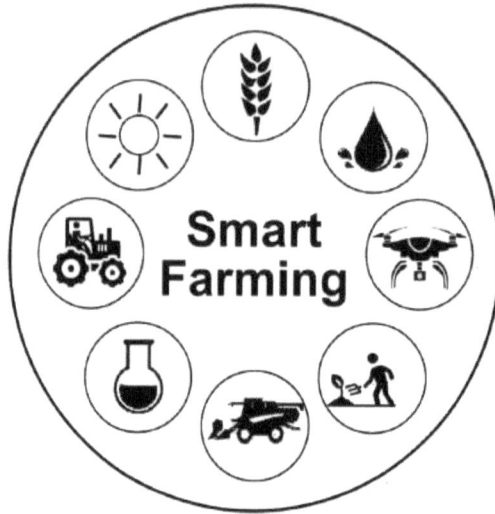

**Fig. (1).** Smart farming tools.

## DIFFERENT SENSORS USED IN AGRICULTURE

A sensor is basically an efficient technology that measures natural conditions like the movement of particles in air, soil, heat transfer, or light and converts these conditions into digital form. There are numerous types of sensors used in precision farming. They are detailed below:

### Optical Sensors:

It measures soil ingredients like clay, sand, and moisture content using various frequencies of light installed in the sensor. The sensors kept on vehicles or drones can collect plant colour data by allowing soil reflectance and then processing the data. These can also evaluate crop conditions. Different plant properties can be measured by different colored light waves.

### Electrochemical Sensors:

Electrochemical sensors are basically designed for soil nutrient detection by collecting chemical data. It uses an ion-selective electrode for the determination of pH and detects the activity of a particular ion in the soil.

### Dielectric Sensors:

The moisture sensors are used to identify the moisture levels in the soil on the farm when there is less vegetative cover. It is used in connection with rain check locations throughout the farm.

## Location Sensors in Agriculture:

With the help of signals received from GPS satellites, these sensors specify the latitude, longitude, and altitude of any farming place.

## Electronic Sensors

This sensor is used to check the functioning of field equipment. The sensor is placed on field equipment and based on the information received in the office computer of the field executive by the captured cellular and satellite data, it provides information regarding the functioning of the equipment.

## Airflow Sensors

An air flow sensor is based on the principle of heat transfer using a silicon chip design. It measures the soil air permeability of a single or multiple locations. This sensor generates the required pressure to push a specified amount of air into the ground at a particular depth.

## Sensors used in Agriculture

This sensor has huge applications in agriculture. It provides necessary information to farmers regarding the various weather parameters required for crop growth. If there is any abnormal change with respect to the parameters observed, then farmers can take early measures to prevent any kind of loss.

## WHAT ARE THE BENEFITS OF SENSORS IN AGRICULTURE?

The biggest challenge in the modern agricultural sector is feeding every mouth with limited resources. The increasing population and lack of proper management have forced the farmers to grow more crops on declining land availability and deteriorating soil. Sensor based applications in agriculture help the farmers constantly monitor the farm conditions and crops/animals in real-time and collect crucial data about the weather, land conditions, crop conditions, livestock, *etc.* The vast amount of data allows farmers to predict changes even before they happen. So, farmers can make informed decisions about their farms and find ways to take necessary measures to solve the problems. The sensor based solutions help to automate the farming process, demand-based fertilizing, irrigation, robot harvesting, resource optimization, and reduced waste and cost management.

## Excelled Efficiency

In the current system of agriculture, many farmers are forced to develop the same crops year after year by distressing and degrading the soil, affecting the ecological systems. Sensor based applications in agriculture provide a decision support system which lets farmers reveal their products and situations in real-time. Additionally, IoT based applications call for primarily based irrigation, fertilizing, and robotic harvesting.

## Expansion

The greenhouses and hydroponic structures, which are IoT enabled, can allow brief food supply chains and capable of feeding people by enabling Smart closed-cycle agricultural structures, permit developing foods essentially everywhere, including rooftops.

## Reduced Sources

The use of IoT is centered on optimizing and using already available sources like water, electricity, and land. In Precision farming, the use of IoT is based on the statistics gathered from various sensors in order to enable the farmers to appropriately allocate resources.

## Cleaning Procedure

The IoT-based precision farming assists farmers in smartly using the irrigation water and electricity, which makes the farming greener. This method helps in getting a cleaner and natural product in comparison to standard methods used in farming.

## Agility

The advantage of IoT-based sensors used in agriculture is the accelerated agility of the processes. It helps in constant monitoring and prediction, which can be helpful to farmers in situations of severe climate change; new skills assist agriculture specialists to protect the plants from adverse climatic conditions.

## Improved Production and Quality

Data driven precise agriculture enables greater and higher production [20] and also maintains quality by using sensors, drones, and many digital applications.

## Monitoring Weather Situations

Weather stations equipped with sensors can acquire climate data and send the same data to the server. Moreover, the climatic data are analyzed with the help of prediction software/programs to receive a premade evaluation [21] for an in-depth forecast which helps in minimizing crop losses.

## Greenhouse Automation

Climate stations can routinely regulate the situations to shape the given climate parameters and offer the most suitable circumstances for every greenhouse. This is also beneficial for the product that makes use of smart agriculture sensors. A smart sprinkler controller can control the irrigation and lighting fixtures remotely.

## Crop Tracking

Sensors used for crop tracking can collect all kinds of statistics like crop fitness, humidity, precipitation, temperature, and different parameters. If there is any deviation, farmers may also discover it in advance using those sensors and take suitable measures. Also, sensors assist in deciding the time of plantation and harvesting.

## Drones

In smart farming sensors, drones have various uses, like soil and crop monitoring and evaluation planting, and pesticide spraying. Drones may be used with one-of-a-kind imaging technology like multispectral, thermal, *etc.*, that could offer the farmers time and site-specific statistics concerning crop fitness, fungal infections, bottlenecks, *etc*. Drones can also discover drier areas in a discipline, and measures can then be taken to irrigate such areas with higher techniques. Precision agriculture presents farmers with such concrete statistics. That permits them to make knowledgeable choices and make better use of their sources.

## APPLICATIONS OF SENSORS IN FARMING

### Applications in Animal and dairy science:

The most important factor in successful livestock performance is determining the number of animals grazing (sustaining) per acre for a specified amount of time. The two major avenues in the livestock production system that involve higher costs are animal feed and proper disease management. Due to this, a farmer tries to minimize the costs by increasing the stocking rate. In animal farming, many researchers tried to solve complex problems like identifying limiting factors, optimizing the nutrient composition of feed, evaluating animal management, and implementing strategies to reduce nutrient excretion. For experimenting with this, a large volume of diverse datasets (big data) is required. Various sensors have the potential to capture large amounts of real-time data effectively, and for processing a large amount of data, a high-end computer is required [22]. Then the sensor can measure the different features and analyze the vast data for interpretation. The sensors also can be used to collect data of weather parameters, air quality and voice signals of animals (Bioacoustics sensor), animal movements and other such animal behavior data. Based on the market needs, the various sensors [23] can be employed for precision milking and feeding systems; infrared thermal imaging sensors; temperature sensor; RFID tags; and accelerometers. These sensors can help the farmers by alerting them to any changes in animal movements, abnormalities in food intake, changes in sleep cycles, or poor air quality in animal shelters. Sensors, in modern farms, have been employed to successfully diagnose the early symptoms like lethargic body movements, slower response times, and decreased activity before the spread of disease. For example, the air sensors in the poultry industry have the feature of detecting the onset of Coccidiosis (an intestinal infection in birds) without any apparent symptoms [24]. Air sensors can detect the changes in volatile organic compounds in the air before normal detection by the farmer or any vet. Sensors scan the environmental data [25] and help the farmers prevent contagious diseases from spreading among animals. All these applications of sensors highlight the importance and relevance of livestock farms in modern farming.

**Artificial Intelligence in sensor-based milking**: The application of automation in milk booths [26] is an important application of sensors in animal husbandry. The AI based milking units check the milk quality and detect any abnormalities present in the product. AI-based technologies [27] are now appearing in the dairy sector as well. There are applications to test milk quality, trace milk to a cow, diagnose and predict animal health, *etc*. In all the above-mentioned areas of application, the technologies are yet to be fully perfected and fully embraced. However, over a ten-year period from today, most of these technologies would become part of the daily lives of agribusinesses and farmers.

**Precision livestock farming**: Sensor based applications have tremendous potential in the latest Dairy farms. These applications help in heat detection, calving, health monitoring, detecting, and making charts of daily routine activities. These sensors can also track any alterations in animal habit, and check the quality of the air in animal shelters, and record information about the reproduction, health, and nutrition status of farm animals.

**Applications of AI in health monitoring of farm animals**: The AI based applications not only help the farmers keep an eye on every cow in the herd but also send alerts to the farmer about changes in the behavior of farm animals and the need for any human intervention (Fig. **2**). Sensor based advanced algorithms also do the surveillance and protect the farms from disease outbreaks. The abnormal animal behaviors of the animal can be detected [28] by several sensors placed at the smart farm and can notify the farmer about the abnormal symptoms of the farm animals (Fig. **3**).

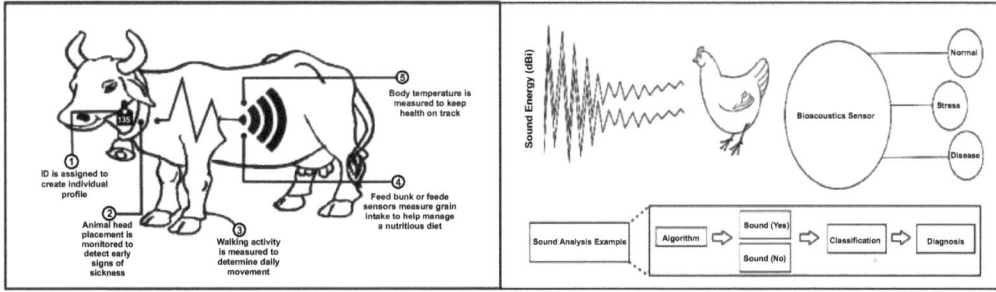

**Fig. (2).** Applications of sensors in animal farming.

**Fig. (3).** AI applications for health monitoring of farm animals.

**Robotic System to Deliver Vaccines:** The Introduction of robotic injection in modern dairy farms is one step forward towards making the dairy farm economically sustainable. Nowadays, robotic systems are widely applicable on dairy farms. The RFID tags placed in the robotic injection system help collect health-related [29, 30] information and the vaccination record of the cow. The automated system directs the injection site to deliver the medication to the cow in case of any requirement. Sensor based big data on animal farming practices stored on a remote server helps inform farmers in case there are any abnormal symptoms. Therefore, sensors based advanced algorithms have the potential to provide a better platform for precise and smart animal farming [31]. Commercial sensors are widely used in modern farms for precise prediction and disease management of livestock through [32] continuous monitoring. However, these sensors are not able to predict the metabolic states of the animals. Advance research/efforts are needed to address those issues, which can make the prediction of farm animals more precise.

**Application of sensor-based technology in agriculture**: Precision Agriculture (PA) is an emerging concept in agricultural management that utilizes several advanced technological applications like Sensors, AI, IOT, and Big Data to optimize the output by reducing environmental impacts. There are huge applications of AI in agriculture, from field preparation to harvesting crops. There are AI based technologies/innovations that are able to correctly assess the field layout and guide a tractor (with or without a driver) to prepare the field in such a manner as to maximize production. These technologies can work without a driver, and accurately assess the lay of the land to maximize land usage. AI-based robotic applications have now made it possible to sow seeds in a straight line with adequate space between them, replacing seed-drills and manual sowing. Pesticides, herbicides, and other agrochemicals are usually sprayed all over the plants and sometimes even randomly around un-affected areas. Current innovations in image-based analytics have made it possible for drone-based sprayers to target infected areas of the plant. AI and IOT based Precision Agriculture uses intelligent control of sensor-based devices connected to wireless networks that are provided with informed decisions by applying advanced algorithms. The different Sensors and algorithms widely used in PA are listed in Table **1**.

**Table 1. Different sensors and algorithms widely used in precision agriculture (PA).**

| S. No. | Name of Sensors | Application in Agriculture |
|---|---|---|
| 1. | Humidity Sensor | It measures water vapor present in the air within an enclosed space |
| 2. | Light Intensity Sensor | Electronic device primarily used to detect the presence of light |
| 3. | $CO_2$ Sensor | Used to measure the carbon dioxide gas concentration |
| 4. | Water Nutrient Level Sensor | Device to detect the water level and provide water nutrient level data in real time |
| 5. | EC and pH Sensor | This device measures the concentration of dissolved salt throughout the growth phase |
| 6. | Soil moisture sensors | These sensors determine the moisture level of the root area of the crops and send the information to the controller part of the irrigation system which will activate the water supply through microcontroller and thereby switch on or off the irrigator pump. This is the most valued technology that has revitalized the sustainable agriculture. |
| 7. | Satellite imagery, cloud machine learning as well as wireless network | The crop growth monitoring and yield prediction |
| 8. | Geo-tagging, geo imaging technologies combined with Bigdata analytics and AI | Crop insurance companies are relying on highly accurate digital data |
| 9. | ZigBee sensor network | Wireless Sensor Networks(WSNs)control the Greenhouse parameters including temperature, pressure, light, humidity, $CO_2$, wind speed *etc.* |

*(Table 1) cont.....*

| S. No. | Name of Sensors | Application in Agriculture |
|---|---|---|
| 10. | Plant biosensors | Management of abiotic stress and diseases of plants using the sensors |

New age AI-based apps have made it possible for farmers to click photos of crops that are infected with diseases and get a probability score [33] of the infected disease on a real-time basis. As in the case of diagnosing diseases, here, too, the farmer has to click photographs of crops to get a probability score of nutrient deficiency on a real-time basis. However, a physical test is recommended for a correct diagnosis. Now it is possible for users to map the lay of the land based on the existing soil nutrients. This is very useful for the application of Urea, DAP, MOP, SSP, and separate doses of N, P & K, which can be appropriately applied on the ground without compromising soil health. Technologies to monitor water release timing, flow, quantum, and speed ensure that water is used most productively based on past data on soil health and soil nutrients. There are technologies that look at environmental factors and soil conditions to predict yields for the coming seasons. There are also image-based technologies that look at images of plantations – palm, rubber, sugarcane, tea, coffee, coconuts, apples, mangoes, *etc*. Machines can now harvest crops more precisely and cleanly, thanks to AI. This is again done by looking at images and cutting crops so as to maximize yield. AI-based technologies now make it possible for farmers and companies to take photographs of commodities with proper display of details on the screen. These results show the count and weight of the commodity with a fair degree of accuracy. Combinations of the above technologies on weather, yield, and quality can be used by insurance companies as reliable data-driven measures to tailor customized insurance policies. Correct processing mixes for both food and feed items can be obtained effortlessly with AI. These mixes may be further customized, based on the animal type, stage of growth, diet preferences, *etc*. The results of the above technologies, especially on yield and quality, can be shared by banks when taking loans. Banks can, in turn, use these data as a basis for funding, given their reliability and retrievability. Some of the agricultural applications are given below:

**Precision Farming by Micro Sensors**: Micro sensors in precision farming are able to scan/monitor their crops for symptoms of any disease or contamination. The technology is an application of machines that can display various information. Farmers can visualize instant alerts about any pests, diseases or others in their field (Fig. **4**).

**Drone Operations**: Crop Monitoring: Cameras and sensors placed inside the drones collect farm information about plants down to a single leaf. Aerial and ground-based drones [34] are used for irrigation, assessing crop health, spraying, monitoring, and field analysis (Fig. **4**).

**Fig. (4).** Sensor applications in Agriculture.

**Predictive Analytics**: The sensor-based applications help the farmers make the best possible crop combinations before the season starts based on the previous year's data. The different sensors provide predictions based on different farm data like pest infestation, weather data, soil and crop usage statistics. It also provides the solution for the exact time of seed sowing, monitoring weather parameters suitable for plant cycles, and alerting in case of any weather hazards. Thus, predictive analytics help the farmers provide early farm-based solutions to minimize environmental and crop losses.

**Digital Monitoring of soil and crop:** Sensor based applications help in the digital monitoring of soil and crops by using image recognition techniques. These solutions can identify the nutrient deficiencies in the soil and can suggest the farmer the time and dosage of fertilizer to be added to the soil. Sensor based devices like Plantix can provide solutions for the detected problems and improve the harvest from the farm.

**Using drones for data collection:** The use of drones [35] in farming helps the farmers capture real time sensor data, including images. It acts as a reliable tool for capturing and storing large amounts of data by video surveillance on farms of all sizes, big or small. After analyzing such big data, it is possible to give solutions to the farmer for improving crop yields on his farm. It helps reduce the time and cost associated with the collection of crop data for any farm.

**Precision farming:** The sensor-based applications act as a guide for modern farmers regarding soil and water management, suitable crops for rotation, precise times for harvest, optimum planting methods, strategies for minimal pest attacks, *etc.* The AI based solutions with Machine and Deep learning algorithms help in getting a customized plan for their lands that can increase food sustainably, grow production and increase revenues without depleting any natural resources on the farm.

**AI powered robots:** The application of robots [36] replaces human labour in farming; by performing manual tasks at a faster rate with large volumes of work. These are trained to do multiple farming operations [37], like controlling the excessive growth of weeds, watering plants, harvesting crops, *etc.* The recently developed Agrobot robots have the ability to work 24/7 which increases efficiency, optimizes the cost of precision, harvests the crops and can identify the ripeness of the crop in the field.

**Sensor based surveillance systems:** Sensor based technology minimizes the risk of crop loss [38] and can generate an alert on farmers smart phones in case of any incidence of insects, for example, locusts, grasshoppers, *etc.*, so that they can take the required precautions. The Plantix app turns the smart phone of the farmer into a mobile crop doctor for accurately detecting pests and serves as a complete guide for crop production and management on any smart farm.

**Applications in Fisheries:** In fisheries, the quality of the water in the pond determines the growth of fish. Water Temperature, pH level, dissolved oxygen, nutrient level, and water salinity are the various water quality parameters that should be maintained properly to increase fish yield [39]. If the appropriate levels of those parameters are not maintained properly, it will disturb the healthy lives of fish. In fish farming, it is really difficult for a fish farmer to continuously monitor and track the activity inside the pond. Humans with years of experience can't do such activities. But the application of different sensors can do this. Biosensors can monitor multiple water quality parameters like pH level, temperature, and oxygen proportion and store the collected data in the form of digital signals [40, 41] for generating alerts to the farmers if the current value exceeds the optimal set point, they can take preventive measures timely (Fig. **5**). Sensor based algorithms can solve the different queries with respect to catches on fishing vessels, seabed morphology and biomass on the seabed. SMARTFISH is a suite of high-tech systems that are newly introduced in the fisheries domain and can help the fishing industry, fisheries managers, and stock assessment scientists in many ways. The development of a smart cage culture [42] management system is based on Artificial Intelligence and Internet of Things (AIoT) to solve relevant problems and promote large-scale cage culture. It can also increase omni-fish production to meet the needs of a growing population. Some fish ranches are kept far from the land and can utilize the IoT to screen water at a distance, which could decrease their expenses. Drones can be utilized in fisheries both above and below the water. Drones can also be used to monitor offshore fish farms. Drones can make a quick survey of the ocean [43] and provide in depth analysis through the use of sensor technology. Drones are also able to collect data on fish stocks and the environment that can be analyzed to create algorithms for further applications in the production of aquaculture. Umitron cells are a smart feeding system for aquaculture, and the world's first real time ocean-based fish application system. Vision-based sensors [44] can analyze swimming habits, height, feeding habits, *etc.*, for cultured animals. Aqua pod uses a waterproof hand-held remote for the free floating of fishes which can accommodate several hundred or several thousand fishes. E-logbooks can record catches (origin and volume) and gear used and can be stored in a database. IOT enables soft computing capable of performing multiple tasks through sensors [45] to produce an in-depth analysis. Some of them are given below:

√ Constant monitoring of DO level at the water body.

√ Constant monitoring of water pH.

√ Digitized Water management.

√ Monitoring fishes' behaviors using digital image processing applications.

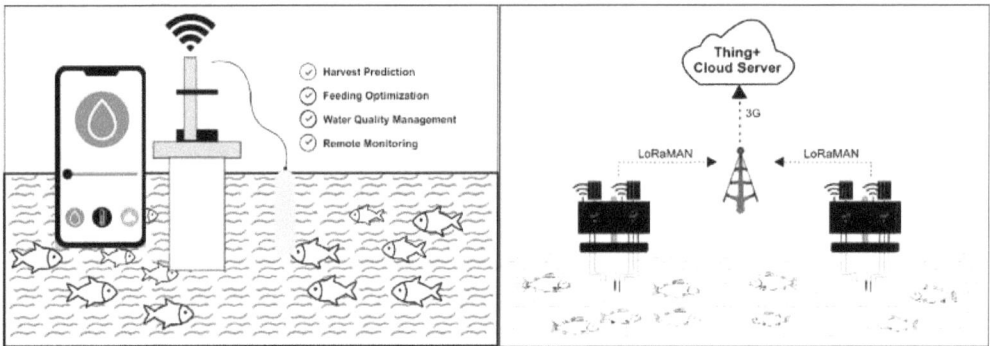

**Fig. (5).** Sensor applications in Fisheries.

But the cost of these sensors is very high, and the application procedure sometimes becomes tedious for the farmers. So, a cost-effective system is required to determine the overall quality of the water effectively and to apply those smart applications in fisheries at a low cost. Some of the water quality parameters are crucial for fish growth and their imbalances cause imbalances in other parameters. From the quantity of some parameters, the condition of others [46] can be assessed. If water temperature, pH and conductivity are balanced, DO is expected to be balanced. A unique aquaculture monitoring system is designed and implemented [47] based on IoT, where both Wi-Fi and the Internet are combined for convenience and to give better results at a lower cost than other available systems. An integrated chip on the computer Raspberry Pi was used in that system for data processing and storing devices using an inbuilt Wi-Fi module. A microcontroller based fish pond [48] using the sensors has also been devised and has immense potential for monitoring those parameters. Under the device, ultrasonic and pH sensors were fixed in the pond system to monitor the water level and water quality. The microcontroller was used to control the whole circuit of the system, the LCD display was connected to the microcontroller, and any fault or issue would be displayed.

Next-generation sensors of various shapes and sizes that sail through the seas are collecting detailed information on everything from salinity, oxygen levels, and temperature to seismic tremors, undersea volcanic eruptions, and even the abundance of schooling fish. This advanced sensor-based technology can help make smarter decisions. Sonar fishing technology, which works in conjunction with smartphones and tablets, detects data and transmits it to the phone in real time for mapping the sea water beds, marking spots, logging water temps and depths, adding lures, species, and making videos in the trip log. Bosch's IoT solutions, Aqua Easy are two popular end-to-end solutions for aquaculture farms (Fig. **6**).

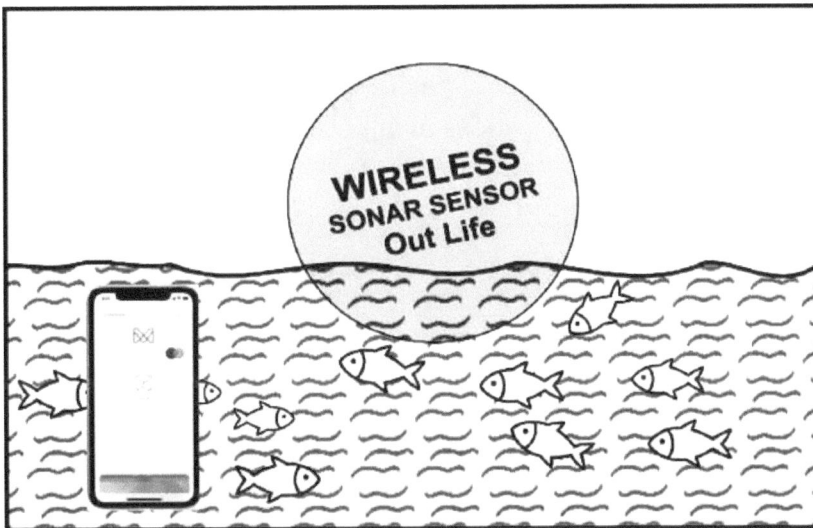

**Fig. (6).** Wireless Sonar Sensor in Fisheries.

## CHALLENGES TO ADOPT SENSOR BASED APPLICATIONS

Some barriers relating to policy relating to adopting sensor based applications were recognized as: (1) issues/policies relating to data governance, including data rights, and (2) universal policy formulation for data regulations, privacy, and transparency of the data, (3) issues related to risk-aversion, (4) less confidence in technology applications, and (5) limited support of academic as well as research institutes in data digitization and (6) higher cost of those sensor based applications in field condition. These issues should be addressed by policymakers to adopt sensor-based applications at a faster rate in agriculture to make the country self-sufficient in agriculture and enhance the economic prosperity of the country.

## CONCLUSION

Sensor based applications in agriculture can enable farmers to automate their farming and have the potential to increase crop yields with better quality while using optimal resources. The big data algorithms have not yet been standardized to collect, process, and present data globally. It is very important to embrace not only these technologies but also start-ups by testing their products in live projects. Many researchers from different corners of the world are trying their level best to propose a prospective low-cost alternative for precise and smart farming that will help with constant monitoring of the farm, making real time decisions, and adjusting quickly to changing environments and conditions. They need to make the product more relevant and customized for their usage, and finally, we need to be generous enough to provide them with business opportunities, however they may be small. Researchers around the globe are dedicatedly involved in improving and correcting the algorithms of AI-based products or solutions for more useful and advanced applications in this sector in the near future, helping the world become self-sufficient in agriculture for the ever-increasing population.

## REFERENCES

[1]     FAO, "Food and agriculture organization of the united nations", *The state offood and agriculture leveraging food systems for inclusive ruraltransformation.*, pp. 109873-109878, 2017.

[2]     Price Policy for Kharif Crops (2017-18). Commission for agricultural costs and prices, department of agriculture, cooperation and farmers welfare, ministry of agriculture and farmers welfare, government of india. Available from: https://cacp.dacnet.nic.in/ViewReports.aspx?Input=2&PageId=39&KeyId=598.

[3]     Price Policy for Rabi Crops (2022-23). Commission for Agriculture Costs and Prices, Department of Agriculture, Cooperation and Farmers Welfare, Ministry of Agriculture and Farmers Welfare, Government of India. Available from: https://cacp.dacnet.nic.in/ViewQuestionare.aspx?Input=2&DocId=1&PageId=42&KeyId=795.

[4]     A. Census, "'All india report on number and area of operational holdings.': Agriculture census division, department of agriculture, co-operation and social welfare, ministry of agriculture and farmers welfare, government of india, pg. 6. agriculture. international conference on robotics and smart manufacturing", *Procedia Comput. Sci.,* vol. 133, pp. 502-509, 2015-16.

[5]     Employment in Agriculture, "The World Bank Report", Available From:www.worldbank.org

[6]     U.M.R. Mogili, and B.B.V.L. Deepak, "Review on application of drone systems in precision", *Procedia Computer Science,* vol. 133, pp. 502-509, 2018.

[7]     G. Shah, A. Shah, and M. Shah, "Panacea of challenges in real-world application of big data analytics in healthcare sector", *Journal of Data, Information and Management,* vol. 1, no. 3-4, pp. 107-116, 2019.
         [http://dx.doi.org/10.1007/s42488-019-00010-1]

[8]   N.S. Ferguson, "Optimization: A paradigm change in nutrition and economic solutions", *Adv. Pork Prod.,* vol. 25, pp. 121-127, 2014.

[9]   A. Baruah, "Artificial intelligence in Indian agriculture: An Indian industry and startup review", Available From:www.emerj.com   https://emerj.com/ai-sector-overviews/artificial-intelligenc-
-in-indian-agriculture-an-industry-and-startup-overview/

[10]  H. Yang, W. Liusheng, and X. Junmin, "Wireless sensor networks for inten- sive irrigated agriculture, consumer communications and networking conference", Las Vegas, Nevada, pp. 197–201, 2007.

[11]  K. Liakos, P. Busato, D. Moshou, S. Pearson, and D. Bochtis, "Machine learning in agriculture: A review", *Sensors,* vol. 18, no. 8, p. 2674, 2018.
[http://dx.doi.org/10.3390/s18082674] [PMID: 30110960]

[12]  R.W. Wall, and B.A. King, "Incorporating plug and play technology into measurement and control systems for irrigation", In: *Management* Ottawa: Canada, 2004, pp. 1-4.

[13]  T. Hemalatha, and B. Sujatha, "Sensor based autonomous field monitoring agriculture robot providing data acquisition & wireless transmission", *Inter. J. Innov. Res. in Comput. Commun. Engi.,* vol. 3, no. 8, pp. 7651-7657, 2015.

[14]  K. Jha, A. Doshi, P. Patel, and M. Shah, "A comprehensive review on automation in agriculture using artificial intelligence", *Artificial Intelligence in Agriculture,* vol. 2, pp. 1-12, 2019.
[http://dx.doi.org/10.1016/j.aiia.2019.05.004]

[15]  K. Kundalia, Y. Patel, and M. Shah, "Multi-label movie genre detection from a Movieposter using knowledge transfer learning", *Augmented Human Research,* vol. 5, no. 1, p. 11, 2020.
[http://dx.doi.org/10.1007/s41133-019-0029-y]

[16]  M. Gandhi, J. Kamdar, and M. Shah, "Preprocessing of non-symmetrical images forEdge detection", *Augmented Human Research,* vol. 5, no. 1, p. 10, 2020.
[http://dx.doi.org/10.1007/s41133-019-0030-5]

[17]  K. Ahir, K. Govani, R. Gajera, and M. Shah, "Application on virtual reality for enhanced education learning, military training and sports", *Augmented Human Research,* vol. 5, no. 1, p. 7, 2020.
[http://dx.doi.org/10.1007/s41133-019-0025-2]

[18]  M.G. Plessen, "Freeform path fitting for the minimization of the number of transitions between headland path and interior lanes within agricultural fields", *Arxiv 1910.12034v1,* pp. 1-7, 2019.

[19]  G.S. Dhaliwal, V. Jindal, and B. Mohindru, "Crop Losses due to insect pests: Global and indian scenario", *Indian J. Entomol.,* vol. 77, no. 2, pp. 165-168, 2015.
[http://dx.doi.org/10.5958/0974-8172.2015.00033.4]

[20]  J. Nagpal, "Digital Agriculture: Farmers in India are using AI to increase crop yields. Microsoft india news center", Available From:www.microsoft.com  https://news.microsoft.com/en-in/features/a-
-agriculture-icrisat-upl-india/

[21]  FE Bureau, "Economic Survey 2017-18: Agriculture- Climate change likely to lower farmer's income by 25%", Available From: https://www.financialexpress.com/budget/economic-survey-2017-
18-agriculture-climate-change-likely-to-lower-farmers-income-by-25-1035560/

[22]  S.G. Matthews, A.L. Miller, T. Plötz, and I. Kyriazakis, "Automated tracking to measure behavioural changes in pigs for health and welfare monitoring", *Sci. Rep.,* vol. 7, no. 1, p. 17582, 2017.
[http://dx.doi.org/10.1038/s41598-017-17451-6] [PMID: 29242594]

[23]  M.M. Rojas-Downing, A.P. Nejadhashemi, T. Harrigan, and S.A. Woznicki, "Climate change and livestock: Impacts, adaptation, and mitigation", *Clim. Risk Manage.,* vol. 16, pp. 145-163, 2017.
[http://dx.doi.org/10.1016/j.crm.2017.02.001]

[24]  F. Borgonovo, V. Ferrante, G. Grilli, R. Pascuzzo, S. Vantini, and M. Guarino, "A data-driven prediction method for an early warning of coccidiosis in intensive livestock systems: A preliminary study", *Animals,* vol. 10, no. 4, p. 747, 2020.

[http://dx.doi.org/10.3390/ani10040747] [PMID: 32344716]

[25] S. Neethirajan, "The role of sensors, big data and machine learning in modern animal farming", *Sens. Biosensing Res.,* vol. 29, p. 100367, 2020.
[http://dx.doi.org/10.1016/j.sbsr.2020.100367]

[26] T. Van Hertem, E. Maltz, A. Antler, C.E.B. Romanini, S. Viazzi, C. Bahr, A. Schlageter-Tello, C. Lokhorst, D. Berckmans, and I. Halachmi, "Lameness detection based on multivariate continuous sensing of milk yield, rumination, and neck activity", *J. Dairy Sci.,* vol. 96, no. 7, pp. 4286-4298, 2013.
[http://dx.doi.org/10.3168/jds.2012-6188] [PMID: 23684042]

[27] K. VanderWaal, R.B. Morrison, C. Neuhauser, C. Vilalta, and A.M. Perez, "Translating big data into smart data for veterinary epidemiology", *Front. Vet. Sci.,* vol. 4, p. 110, 2017.
[http://dx.doi.org/10.3389/fvets.2017.00110] [PMID: 28770216]

[28] C. Pomar, and A. Remus, "Precision pig feeding: A breakthrough toward sustainability", *Anim. Front.,* vol. 9, no. 2, pp. 52-59, 2019.
[http://dx.doi.org/10.1093/af/vfz006] [PMID: 32002251]

[29] T. Debnath, S. Bera, S. Deb, P. Pal, N. Debbarma, and A. Haldar, "Application of radio frequency based digital thermometer for real-time monitoring of dairy cattle rectal temperature", *Vet. World,* vol. 10, no. 9, pp. 1052-1056, 2017.
[http://dx.doi.org/10.14202/vetworld.2017.1052-1056] [PMID: 29062193]

[30] T. Debnath, S. Bera, S. Deb, P. Pal, N. Debbarma, D.D. Choudhury, and A. Haldar, "Real-time monitoring of peripheral body temperature using non-invasive, self-powered, sensor based radio-frequency device in goats (capra hircus)", *Small Rumin. Res.,* vol. 144, pp. 135-139, 2016.
[http://dx.doi.org/10.1016/j.smallrumres.2016.09.007]

[31] E. Fernández-Carrión, M. Martínez-Avilés, B. Ivorra, B. Martínez-López, Á.M. Ramos, and J.M. Sánchez-Vizcaíno, "Motion-based video monitoring for early detection of livestock diseases: The case of African swine fever", *PLoS One,* vol. 12, no. 9, p. e0183793, 2017.
[http://dx.doi.org/10.1371/journal.pone.0183793] [PMID: 28877181]

[32] M. Taneja, J. Byabazaire, N. Jalodia, A. Davy, C. Olariu, and P. Malone, "Machine learning based fog computing assisted data-driven approach for early lameness detection in dairy cattle", *Comput. Electron. Agric.,* vol. 171, p. 105286, 2020.
[http://dx.doi.org/10.1016/j.compag.2020.105286]

[33] Available From:https://news.panasonic.com/global/stories/2018/57801.html

[34] T. Fictchett, "Netafim drip irrigation success story' western farm press", Available From:https://www.netafimusa.com/wp-content/uploads/2016/08/Alfalfa-Success-Maddox-2013.pdf

[35] D.G. Panpatte, "Artificial intelligence in agriculture: An emerging era of research", In: *Anand Agricultural University*, 2018, pp. 1-8.

[36] K. Sennaar, "AI in Agriculture present applications and impact", Available From:www.emerj.com https://emerj.com/ai-sector-overviews/ai-agriculture-present-applications-impact/

[37] S.K. Verma, S.B. Singh, R.N. Meena, S.K. Prasad, and R.S. Meena, "A review of weed management in India: The need of new directions for sustainable agriculture", *Bioscan,* vol. 10, no. 1, pp. 253-263, 2015.

[38] J. Stoltzfus, "The 6 most amazing AI advances in agriculture", Available From: www.techopedia.com/https://www.techopedia.com/the-6-most-amazing-ai-advances-in-agriculture/2/33177

[39] S.U. Kiruthika, S.R. Kanaga, and R. Jaichandran, "IOT based automation of fish farming", *J. Adv. Res. in Dyna. Cont. Sys.,* vol. 9, no. 1, 2017.

[40] C.M. Fourie, D.V. Bhatt, B.J. Silva, A. Kumar, and G.P. Hancke, "A solar-powered fish pond

management system for fish farmng conservation", *2017 IEEE 26th International Symposium on Industrial Electronics (ISIE),* Edinburgh, UK, pp. 2021-2026, 2017.
[http://dx.doi.org/10.1109/ISIE.2017.8001565]

[41]   F.E. Idachaba, J.O. Olowoleni, A.E. Ibhaze, and O.O. Oni, "IoT enabled real-time fishpond management system", *Proceedings of the World Congress on Engineering and Computer Science,* San Francisco, USA WCECS Vol I 2017.

[42]   C.C. Chang, J.H. Wang, J.L. Wu, Y.Z. Hsieh, T.D. Wu, S.C. Cheng, C-C. Chang, J-G. Juang, C-H. Liou, T-H. Hsu, Y-S. Huang, C-T. Huang, C-C. Lin, Y-T. Peng, R-J. Huang, J-Y. Jhang, Y-H. Liao, and C-Y. Lin, "Applying artificial intelligence (AI) techniques to implement a practical smart cage aquaculture management system", *J. Med. Biol. Eng.,* vol. 41, pp. 652-658, 2021.
[http://dx.doi.org/10.1007/s40846-021-00621-3]

[43]   F. Akhter, H.R. Siddiquei, M.E.E. Alahi, and S.C. Mukhopadhyay, "Recent advancement of the sensors for monitoring the water quality parameters in smart fisheries farming", *Computers,* vol. 10, no. 3, p. 26, 2021.
[http://dx.doi.org/10.3390/computers10030026]

[44]   M. Irimia, "Five ways agriculture could benefit from artificial intelligence. AI for the Enterprise", Available From:https://www.ibm.com/blogs/watson/2016/12/five-ways-agriculture-benefit-artifiial-intelligence/

[45]   S. Saha, R.H. Rajib, and S. Kabir, "IoT based automated fish farm aquaculture monitoring system", *2018 International Conference on Innovations in Science, Engineering and Technology (ICISET),* Chittagong, Bangladesh, pp. 201-206, 2018.
[http://dx.doi.org/10.1109/ICISET.2018.8745543]

[46]   C.H.E.N. Yongqiang, L.I. Shaofang, L. Hongmei, T. Pin, and C.H.E.N. Yilin, "Application of intelligent technology in animal husbandry and aquaculture industry", *2019 14th International Conference on Computer Science & Education (ICCSE),* Toronto, ON, Canada, pp. 335-339, 2019.
[http://dx.doi.org/10.1109/ICCSE.2019.8845527]

[47]   R. Sharma, "Artificial intelligence in agriculture: A review", *2021 5th International Conference on Intelligent Computing and Control Systems (ICICCS)* Madurai, India, pp. 937-942, 2021.
[http://dx.doi.org/10.1109/ICICCS51141.2021.9432187]

[48]   S. Sivakumar, and V. Ramya, "An intuitive remote monitoring framework for water quality in fish pond using cloud computing", *IOP Conf. Series Mater. Sci. Eng.,* vol. 1085, no. 1, p. 012037, 2021.
[http://dx.doi.org/10.1088/1757-899X/1085/1/012037]

# CHAPTER 10

# An Overview of Building a Global Data Area on the Web for Farming

**R. Sapna**[1,*], **Ravva Akash Guptha**[2], **Paritala Venkateswara Rao**[2] and **Raavi Sai Pranay**[2]

[1] *Department of Information Technology, Manipal Institute of Technology, Bangalore, Manipal Academy for Higher Education, Manipal, India*

[2] *School of Computer Science and Engineering, Presidency University, Bangalore, India*

**Abstract:** A collection of best practices for publishing as well as connecting structured data on the Web is referred to as "Linked Data." In the past ten years, the practice of Linked Open Data has significantly increased the structured data available on the Web. This structured data provides machine-readable descriptions of real-world items, opening hitherto unheard-of opportunities for research around natural language processing. With the aid of technology, the agriculture sector advanced significantly, becoming more data-driven and intelligent. Studies on how these data can be used, for what kinds of tasks, and to what degree they can be helpful for agricultural tasks, however, are lacking. Our paper puts an effort into providing an overview of concepts and technologies in the Semantic Web and their applicability to farming.

**Keywords:** Web 3.0, Ontology, Semantic web stack, Farming, Farming, Good Health, Well-being.

## INTRODUCTION

There are a substantial number of URLs on the World Wide Web (consequently insinuated as the Web), making it quick to create a dissemination system. Over the last decade, Internet use has extended to 4.9 billion people, making it the world's fundamental strategy for correspondence [1]. The development of the plan of the Internet is creating at a reliably growing rate. Being awake to date with creative examples sets out new entryways for relationships as well as hardships. Several studies use text as a source of data to supplement their knowledge bases [2]. As an enabling specialist for mechanical progression, the Web has been created in its original way. From the get-go, there were static illuminating compo-

---

* **Corresponding author R. Sapna:** Department of Information Technology, Manipal Institute of Technology, Bangalore, Manipal Academy for Higher Education, Manipal, India; E-mail: sapna.aradhya@gmail.com

S. Gowrishankar, Hamidah Ibrahim, A. Veena, K. P. Asha Rani & A. H. Srinivasa (Eds.)

nents in Web 1.0, which were made from a dormant experience to a smart one in Web 2.0. The accompanying time of Web headway, Web 3.0, is at this point in the works. New entryways will ascend out of the headway of the Web. An all-out revamping of the Internet and IT establishment is normal for Web 3.0. Affiliations ought to start preparing for the changes, if not, they cannot address client issues, gain from emerging examples, and promptly make the most of new possibilities. Affiliations need to grasp the impact of advancement on business exercises before rushing to totally capitalize on these important entryways. The justification behind this study is to conclude the potential perils that affiliation could face while interfacing with Web 3.0 advancements (insinuated as Web 3.0). This study means to give heads, board people, IT specialists, and information chiefs information about the perils related to Web 3.0, and give ideas on the most capable strategy to direct these threats to an OK level. The motivation behind this study isn't to portray exhaustively the advances in basic Web 3.0, but rather to give an understanding of potential dangers emerging from the utilization of these innovations.

## ABOUT THE WEB AND ITS HISTORY

Lately, there has been an inclination to have rendition numbers for the World Wide Web. The most recent rendition everybody is discussing is Web 3.0. There are still a lot of conflicts about which innovations drive the third era of Websites and Web applications.

Even though the term Web 3.0 is yet equivocal, truly, the majority of the advancements and standards specialists are discussing are not new.

The web is a vehicle for getting to and sharing data in our cutting-edge society. In considering human exercises like recreation, portability, and well-being, Web 3.0 should go past the customary web by incorporating ways of communicating with genuine articles that have commonly not been viewed as registering substances, for example, vehicles and medical care gadgets.

To completely get a handle on what Web 3.0 is about, we want to look at the qualities that portrayed the previous variants.

There were basically static pages on the internet between 1991 and 2004, meaning you just loaded the page, and it showed some content. Posts could not be viewed or logged into, and no analytics were observed; it's called Web 1.0. Then, there is Web 2.0, which emerged around 2004. During this time, the web underwent significant development, but the most significant change may have been the intuitiveness of the web, which meant that, in addition to receiving information from website pages, they also started receiving information directly from us as we used Facebook as well as YouTube and conducted searches on Google. Eventually, they understood that they could bundle up every one of the information they had gathered about us so they could serve us better and we'd remain on their sites longer. The next is Web 3.0, which is the advancement of the web, most likely using blockchain innovation and the instruments of decentralization. The overview of the Evolution of Web is shown in Fig. (**1**).

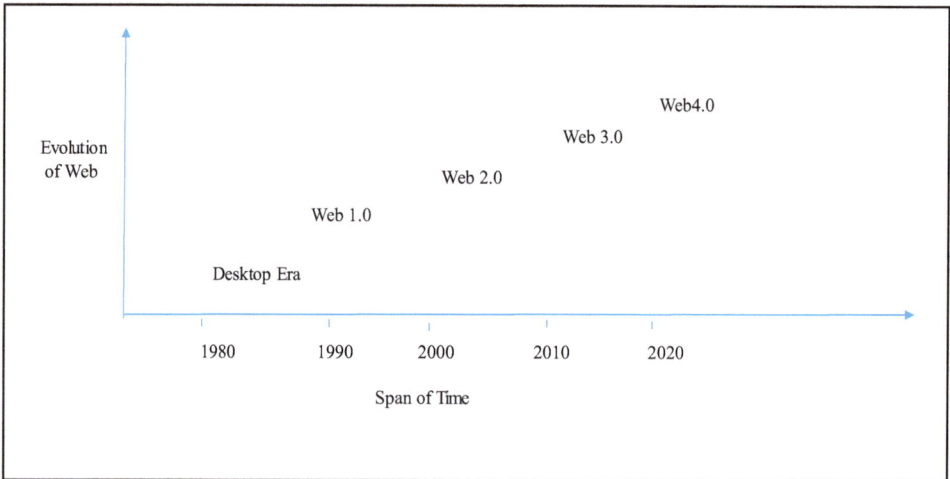

**Fig. (1).** General Methodology employed in Linked Data [7].

## First Era (Web 1.0)

Much of the time, this period is compared to the early long stretches of the World Wide Web. Initially, the Web was an assortment of (for the most part) static pages that held data and content made by organizations or different associations that planned the pages or locales. There is an agreement that during this period, making content was an undertaking performed by specialists, and having individual pages or posting individual substance on the Web was not simple. Thus, clients were only data shoppers in the primary adaptation. Furthermore, pages were isolated snippets of data. To see data from various sources, one should move starting with one site and then onto the next, neglecting to focus on the data contained in past locales.

## Second Era (Web 2.0)

In the second era of the Web, the client is engaged to make content and take part in the Web to uncover themselves and associate with others. Here, the accentuation is on advancements that work with joint effort, for example, informal communities, RSS channels, sites, and content-distributing administrations (pictures, text, and video). The devices are remarkably simple to utilize, which permits anybody to distribute various sorts of content. It has become more straightforward to make dynamic and intelligent Web 2.0 pages by consolidating data from numerous sources into one single page, given the interests of the clients.

## Third Era (WEB 3.0)

The arrangement of innovative web advances that will be characterized as Web 3.0 is yet indistinct. Significant IT specialists in the field have various thoughts regarding what the future web will resemble. As one would expect, such dreams are one-sided by the kind of business they direct or the way they get things done.

The master who was liable for the underpinning of the World Wide Web, Tim Berners-Lee, instituted the expression "Semantic Web". He advocates changing the Web into an enormous information base where modern and complex questions can be made [3, 4].

On the off chance that we need a more insightful Web, with instruments that permit us to find the data when we want it, getting a handle on the information is essential. The data should be organized so machines can peruse and grasp it as well as people, without equivocalness. Along these lines, we will want to make smart specialists that can follow up for our benefit and recover the data we want, as we would assume we were following up on our own.

It is anything but another idea. It has been around without truly having an effect. The semantic web faces two fundamental difficulties in its development. The first is the work to interface existing substance to semantic importance by carrying out some type of metadata. General clients who are not specialists in rationale should have the option to make the machine-reasonable semantic substance. This issue could be addressed by giving a simple method for increasing substance with semantic labels, which ought to be done consequently and for a minimal price or, in any event, free of charge.

The future Web will be a consequence of unavoidable and universal processing situations. The Web will not be lost as far as we might be concerned, however, it will advance and stretch out to become ubiquitous in our day-to-day routines. Admittance to a wide range of data on the Web will be as regular as seeing the time on our watch. We will want to counsel the data, and now and again, it will likewise influence our current circumstance without us, in any event, acknowledging it. The virtuality of web applications and administrations will converge with the truth of our lives.

## SEMANTIC WEB STACK

The Semantic Web Stack, likewise referred to as Semantic Web Cake or Semantic Web Layer Cake, delineates Semantic Web engineering. A global norms body, the World Wide Web Consortium (W3C), drives the Semantic Web drive.

The standard advances conventional information and anticipates the Internet. By connecting with the possibility of semantic substance in website page pages, the Semantic Web targets changing over the consistent web, overwhelmed by unstructured and semi-facilitated reports into a "web of information". The Semantic Web stack fosters the W3C's Asset Depiction System.

The Semantic Web Stack is a framework for the organized development of vernaculars, where each layer makes use of the potential of other layers. Semantic Web Network portrays how advancements that are normalized for Semantic Web work to make the Semantic Web conceivable. Besides, it depicts how the Semantic Web is an expansion (not a substitution) of the traditional hypertext web.

The outline was made by Tim Berners-Lee. The stack is, at this point, concreted to create the layers. (Note: A shrewd visit creating the Semantic Web stack was brought about at the 2009 Worldwide Semantic Web).

### Semantic Web Technologies

As demonstrated inside the semantic web stack, the ensuing dialects or innovations are utilized to make the semantic web. The period from the lower part of the stack, as a lot as an OWL, is right now normalized and standard to build semantic net projects. It's some distance by and by, at this point, not perfect how the apex of the stack will be completed. All components of the stack should be executed to accomplish the total dreams of the semantic web.

### Hypertext Web Technologies

The base layers comprise deep-rooted advancements from the hypertext web that, without change, give establishment of the semantic web.

For the top-layer usefulness of the Semantic Web to be provable, Internationalized Resource Identifiers (IRI) are utilized for exceptionally recognizing assets on the semantic web. IRIs are speculations of URIs.

A semantic web can address records in various human dialects by using Unicode and, furthermore, address reports in different dialects by utilizing Semantic Web.

In XML, archives formed from semi-organized information are markup dialects. In the semantic web, semi-organized information has meaning (semantics).

To involve markup from a few sources in the Semantic Web, XML Namespaces are required. The semantic Web is tied in with associating information, and to this end, various sources should be alluded to in one report.

## Standardized Semantic Web Technologies

The major international body for setting standards for the WWW and creating recommendations for the SW is the World Wide Web Consortium (W3C) [5]. The key SW standards are Resource Description Framework (RDF) and OWL (Web Ontology Language) [2].

An RDF is a method for depicting information about assets as purported significantly increases. When joined with the semantic web, it is at times alluded to as the Giant Global Graph. In the RDF Schema (RDFS) essential jargon is characterized. For instance, making various leveled classes and properties is conceivable.

In the following, we consider a few examples for a clear grasp of the RDF triple form. Similar triple forms are depicted in a tabular form in Table **1**.

**Statement 1:** The title (http://purl.org/dc/elements/1.1/title) of the Web page http://www.w3.org/RDF/DesignIssues/Overview.html is "Design Issues for the World Wide Web"

**The subject of the above statement:** http://www.w3.org/RDF/DesignIssues/ Overview.html (URI)

**The predicate is:** http://purl.org/dc/elements/1.1/title (URI)

**The object is:** "Design Issues for the World Wide Web" (Literal)

**Statement 2:** The maker (http://xmlns.com/foaf/0.1/maker) of the web page http://www.w3.org/RDF/DesignIssues/Overview.html  is  https://www.w3.org/ People/Berners-Lee/card#i

**The subject of the above statement:** http://www.w3.org/RDF/DesignIssues /Overview.html (URI)

**The predicate is:** http://xmlns.com/foaf/0.1/maker (URI)

**The object is:** https://www.w3.org/People/Berners-Lee/card#i (URI)

**Table 1. RDF Triples.**

| Subject | Predicate | Object |
|---|---|---|
| http://www.w3.org/RDF/Desig nIssues/Overview.html | http://purl.org/dc/elements /1.1/title | "Design Issues for the World Wide Web" |
| http://www.w3.org/RDF/Desig nIssues/Overview.html | http://xmlns.com/foaf/0.1/ maker | https://www.w3.org/People/ Berners-Lee/card#i |
| https://www.w3.org/People/Ber ners-Lee/card#i | http://www.w3.org/1999/ 02/22-rdf-syntax-ns#type | http://www.w3.org/2000/10 /swap/pim/contact#Male |
| https://www.w3.org/People/Ber ners-Lee/card#i | http://xmlns.com/foaf/0.1/ family_name | "Berners-Lee" |
| https://www.w3.org/People/Ber ners-Lee/card#i | http://xmlns.com/foaf/0.1/ givenname | "Timothy" |

With Web Ontology Language (OWL), it is feasible to communicate the semantics of RDF proclamations of a further developed nature that incorporate cardinality, limitations of values, and transitivity, as well as different qualities. Table **2** lists out few key ontologies which are generally used.

**Table 2. Key Ontologies for Enrichment.**

| Subject | Predicate | Object |
|---|---|---|
| http://www.w3.org/RDF/Desig nIssues/Overview.html | http://purl.org/dc/elements /1.1/title | "Design Issues for the World Wide Web" |
| http://www.w3.org/RDF/Desig nIssues/Overview.html | http://xmlns.com/foaf/0.1/ maker | https://www.w3.org/People/ Berners-Lee/card#i |
| https://www.w3.org/People/Ber ners-Lee/card#i | http://www.w3.org/1999/ 02/22-rdf-syntax-ns#type | http://www.w3.org/2000/10 /swap/pim/contact#Male |
| https://www.w3.org/People/Ber ners-Lee/card#i | http://xmlns.com/foaf/0.1/ family_name | "Berners-Lee" |
| https://www.w3.org/People/Ber ners-Lee/card#i | http://xmlns.com/foaf/0.1/ givenname | "Timothy" |

The questioning language is important to recover data from semantic web applications utilizing RDF-based information (*i.e.*, proclamations including OWL and RDFS). SPARQL is an RDF question language that can be utilized to inquire about any RDF-based information.

In RIF, rules can be traded. It is critical to help, for instance, the portrayal of relations that cannot be made sense of straightforwardly through the depiction rationale in OWL. The general methogologies used in Linked data are shown in Fig. (**2**). It shows the details of how the raw structured data can be linked and displayed as per the requirement [6].

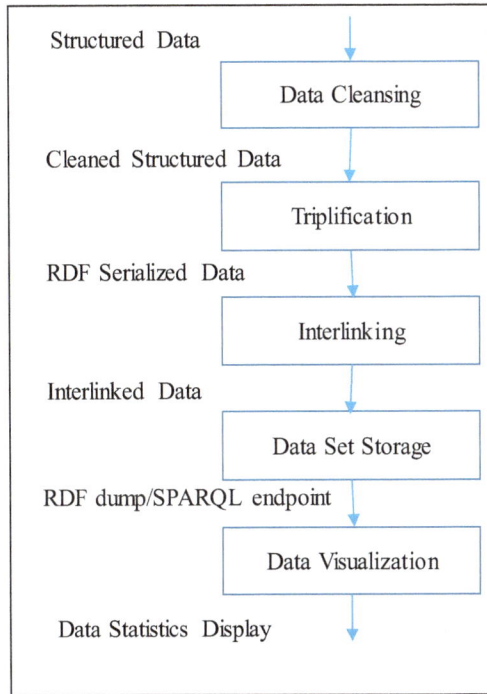

**Fig. (2).** General Methodology employed in Linked Data [6].

## Unrealized Semantic Web Technologies

Those at the top layer contain advances that poor people have yet been normalized or thoughts that should be tried to understand the Semantic Web.

By carefully marking RDF proclamations, you can guarantee that they come from confided-in sources. Cryptography is significant for guaranteeing that such assertions are coming from confided-in sources.

Clients will want to believe determined proclamations by confirming that the premises begin from confided-in sources and by involving formal rationale in the determination of new data. Finally, the User Interface will empower semantic web applications to work for people.

## Machine Learning on Semantic Web

Machine learning will face a variety of new opportunities and challenges because of data in Semantic Web forms. By utilizing patterns in the data and being resistant to some of the inherent issues with Semantic Web data, such as contradictory information and non-stationarity, machine learning enhances ontological background knowledge. A general challenge with machine learning is that the problem of missing information needs to be carefully addressed in learning, especially if the statistical units are chosen or the likelihood that a feature is missing depends on the features of interest, which is typical in many-to-many connections [7].

The vital goal of semantic data is to empower computers to perform more useful work in a better approach and then to further develop systems that can provide dependable interactions over the network. LOD is regarded as a set of best practices that encourages enterprises to publish and interlink their data on the WWW using existing ontologies. In simpler terms, the concept of linked data is a way of closely collaborating data on the Web by links.

To be a part of the linked data, a resource in the source data set should have an RDF link with the resource of another data source.

Using the RDF links, we can traverse through the graph to discover additional details about the resources. For the construction of an RDFS graph, it is required to explore graph structure and Knowledge Base semantics. Nodes that are important can be noted with respect to relevance. From this important node, we maximize its importance to decide upon which connected edge to be chosen. Ontology mining should involve all the accomplishments that let the discovery of hidden knowledge from associated ontological knowledge bases [8]. All these can be thrived by means of machine learning methods.

The Semantic Web has its original motivations to surge automation in processing information that is Web-based and to advance the interoperability of Web-based information systems. Adding machine learning concepts to infer hidden knowledge in the linked data will give us a broad spectrum of data semantics. This is an approach to making data on the Web not just machine-readable but also machine understandable [9, 10].

The application of reasoning on learned data can remove anomalies and infer hidden knowledge, which will be an add-on to the semantic Web. The literature survey elaborates on the reasoning required for obtaining additional knowledge. This can be used to suggest new inferences to the ontology.

The application of machine learning to the Semantic Web will increase its ability enormously. There is flourishing progress in both giant fields of LOD and machine learning. When they are blended in the appropriate way, we can achieve even more progress to get more information about an entity. Unifying these two giant fields will keep RDF as the standard format of data representation in Machine learning.

## Semantic Web and Agriculture

Because of recent advances in the Internet of Things (IoT), it is now feasible to process a huge number of sensor data streams utilising various large-scale IoT platforms. These IoT frameworks are used to collect, process, and interpret data streams in real time, allowing for the deployment of smart solutions that give decision assistance. Interoperability is a difficult challenge for IoT developers all over the world. This is because the IoT devices currently in use use a range of data formats, protocols, and technologies to function. Interoperability tools are currently limited due to the lack of standardised criteria for IoT applications. Semantic Web Technology can be a solution to this problem. The RDF can be employed to give data semantic functionality [11]. The semantic web is a tool that is neglected in agricultural science. Regardless of the fact that there are many agricultural-specific resources, there are few instances in past studies that use semantic resources to tackle agricultural problems [12].

## The Semantic Web Technology for Agriculture

The use of semantic web technologies has grown in popularity as a means of giving unstructured data interpretation. Compared to challenges in related fields, they have not been used to agricultural issues very often. A growing amount of raw data is being produced by agriculture, particularly specifically precision farming, through resources including soil sensing devices, drones, and local weather stations. Because raw data is meaningless as well as isolated, it may not be of much use to farmers. Data's usefulness is determined by its context and meaning as well as by how it is combined with data gathered from other sources. By offering standard data transfer formats and data description languages, semantic web technology may give context and meaning to data in addition to enabling its aggregation.

The agricultural field now has access to a wide range of significant semantic resources and data exchange standards because of work by organisations like the Food and Agriculture Organization of the United Nations (FAO). Activities in agriculture rely on a network of interlinked knowledge. For instance, a crop's production depends not only on its species but also on the soil's nature, the weather in the area, the presence of pests, and the extent of weed infestations. It is improbable that any one knowledge base will have all this data at a fine enough level of detail to be helpful to a single farmer. Utilizing semantic web technologies, that are far more to exist as discrete information stores on each of these domains than as a single monolithic knowledge base, it is possible to counteract this dependence on knowledge from other fields. Using semantic web technologies, disparate knowledge repositories can be synchronised. It will be possible to query it as a broader, interconnected body of information owing to this approach [12].

Data is increasingly important in precision agriculture. Semantic web technologies can offer a uniform representation of data obtained across non-real-time sources like producer and retail systems in addition to real-time sensors. One of the key challenges in precision agriculture is data integration [13, 14]. As a result, semantic web technologies are beginning to take the limelight.

**The Semantic Resources for Agriculture**

The adoption of semantic web technologies is dependent upon the availability of existing semantic resources. By integrating cutting-edge technology and data-driven approaches, semantic agriculture is revolutionizing farming practices, playing a pivotal role in advancing both good health and overall well-being within agricultural communities [15-17]. A controlled vocabulary is basically a list of concepts or words that have been carefully chosen for use in a particular domain. In contrast, a taxonomy collects controlled vocabulary in a structured, hierarchical structure that resembles a tree. Both the entity's parents and the edge that connects them both contain meta-data about the entity. Due to a deliberate but fragmented effort to provide semantic resources for agriculture over many national agencies, agriculture is well serviced by easily available materials [18-23]. Agriculture in general [24] and specific sub-domains of agriculture make up the two primary categories of semantic resources.

The Food and Agriculture Organization (FAO) established AGROVOC, the biggest and most extensive semantic resource. This controlled vocabulary has 40,000 terms as well as 35,000 concepts in it. AGROVOC includes information on food, nutrition, fisheries, forestry, and the environmental field in addition to vocabulary and concepts related to agriculture. English, Arabic, and Chinese are just a few of the 27 languages available in this. The Chinese Agriculture Thesaurus [25], the Agricultural Thesaurus from the National Agricultural Library [26], as well as resources in related fields like the Environmental Applications Reference Thesaurus and more general, unrelated resources like DBPedia [25], have all been aligned with AGROVOC since it supports the Linked Open Data Schema (LOS).

There are 128.253 agricultural terms in English and Spanish in the National Agricultural Library's Agricultural Thesaurus (NALT). 17 subject headings, comprising Farms and Farming Systems and Rural and Agricultural Sociology, are included in the thesaurus. The thesaurus incorporates AGROVOC as well as additional semantic resources like Aquatic Sciences and Fisheries Abstracts, and it supports LOS (ASFA).

The Chinese Agricultural Thesaurus (CAT) [27], which is similarly connected to AGROVOC, includes 40 categories like crop classification and 63,000 ideas like Legume Crop and Azuki Bean Mosaic Virus. It also supports LOS and has thus been aligned not just with AGROVOC, but also with other resources including EUROVOC and LCSH.

The integration of AGROVOC with NALT and CAT shows the original goal of a data-linked network. AGROVOC is not only connected with NALT and CAT, but also tacitly to EUROVOC as well as ASFA *via* CAT and NALT. As a result, the aligned resources can be accessed as a single data source. This has consequences for agriculture since specialised resource developers can extend the key resources that utilise LOS, like AGROVOC.

In addition, frameworks for assessing the quality of linked data in this field can help to boost the research. Quantum computing has the potential to further change agriculture by increasing yields and decreasing overhead expenses.

## CONCLUSION

In the upcoming age of the Internet, administrations are totally based on the clients' ideas. Moreover, the vision for Web 3.0 imagines what is going on where such universal innovations make a consistent combination of real and virtual conditions permitting people and machines to collaborate continuously. Agriculture contains a large number of semantic resources. These are hastily assembled vocabularies, ontologies, and thesauri. Large and diverse resources have been connected together using linked data hubs or shared vocabularies. Furthermore, these resources are freely available and accessible to the general public. As a result, it is unexpected that the adaption of these technologies in practices is restricted, particularly when compared to complementary domains like bio-medicine, education and so on. This study is intended to promote further research into the use of semantic web technologies in agriculture.

## REFERENCES

[1]    Available From: https://worldpopulationreview.com/country-rankings/internet-users-by-country (Accessed on August 19 2022).

[2]    P. Bloem, and G.K. De Vries, *Machine learning on linked data, a position paper.* Linked Data for Knowledge Discovery, 2014, pp. 15-19.

[3]    G. Stix, J. Hendler, and O. Lassila, "Saying yes to no", *Sci. Am.,* vol. 285, no. 5, pp. 34-43, 2001.
       [http://dx.doi.org/10.1038/scientificamerican1101-34] [PMID: 11681174]

[4]    B. Berendt, A. Hotho, and G. Stumme, "Towards semantic web mining", *International Semantic Web Conference* pp. 264–278, 2002.
       [http://dx.doi.org/10.1007/3-540-48005-6_21]

[5]    "W3C: World wide web consortium,", Available From:http://www.w3.org/ (Accessed on August 19 2022).

[6]     S.S. Rao, and A. Nayak, "LinkED: A novel methodology for publishing linked enterprise data", *CIT J. Comput. Inf. Technol.*, vol. 25, no. 3, pp. 191-209, 2017.
[http://dx.doi.org/10.20532/cit.2017.1003477]

[7]     R. Sapna, H.G. Monikarani, and S. Mishra, "Linked data through the lens of machine learning: An enterprise view", *2019 IEEE International Conference on Electrical, Computer and Communication Technologies (ICECCT)*, Coimbatore, India, pp. 1-6, 2019.
[http://dx.doi.org/10.1109/ICECCT.2019.8869283]

[8]     Berners-Lee, "Linked Data Rules", Available From: https://pitt.libguides.com/metadatadiscovery/ linked-data (Accessed on August 19 2022).

[9]     J. Aasman, "Transmuting information to knowledge with an enterprise knowledge graph", *IT Prof.*, vol. 19, no. 6, pp. 44-51, 2017.
[http://dx.doi.org/10.1109/MITP.2017.4241469]

[10]    S. Villata, N. Delaforge, F. Gandon, and A. Gyrard, "Social semantic web access control", Available From:http://sdow.semanticWeb.org/2011/pub/sdow2011_paper_5.pdf(Accessed on August 19 2022).

[11]    HG, M.R. and S. Mishra. An investigative study on the quality aspects of linked open data. In *Proceedings of the 2018 International Conference on Cloud Computing and Internet of Things*, 2018, pp. 33-39.
[http://dx.doi.org/10.1016/j.inpa.2019.02.001]

[12]    N. Zhang, M. Wang, and N. Wang, "Precision agriculture : A worldwide overview", *Comput. Electron. Agric.*, vol. 36, no. 2-3, pp. 113-132, 2002.
[http://dx.doi.org/10.1016/S0168-1699(02)00096-0]

[13]    B. Drury, R. Fernandes, M.F. Moura, and A. de Andrade Lopes, "A survey of semantic web technology for agriculture", *Inf. Process. Agric.*, vol. 6, no. 4, pp. 487-501, 2019.

[14]    "TopQuadrant, controlled vocabularies, taxonomies, and thesauruses (and ontologies), technical report, topquadrant", Available From: http://www.topquadrant.com/docs/ whitepapers/cvtaxthes.pdf (Accessed on August 19 2022).

[15]    F. Celli, T. Malapela, K. Wegner, I. Subirats, E. Kokoliou, and J. Keizer, "AGRIS: Providing access to agricultural research data exploiting open data on the web", *F1000 Res.*, vol. 4, p. 110, 2015.
[http://dx.doi.org/10.12688/f1000research.6354.1] [PMID: 26339471]

[16]    G.O. Ekuobase, and E.P. Ebietomere, "Ontology for alleviating poverty among farmers in nigeria", *Proceedings of the 10th International Conference on Informatics and Systems.* pp. 28–34 ,2016.
[http://dx.doi.org/10.1145/2908446.2908465]

[17]    G.G. Aastrand, M. Kiesel, M. Abufouda, and A. Schro¨der, *Semantic integration through linked data in the igreen project.*, G.I.L. Jahrestagung, Ed., , 2012, pp. 107-110.

[18]    S. Joo, S. Koide, H. Takeda, D. Horyu, A. Takezaki, and T. Yoshida, *Agriculture activity ontology: An ontology for core vocabulary of agriculture activity.* In ISWC, 2016.

[19]    S. Iti, "Integrated taxonomic information system, Website", Available From:http://www.itis.gov/ (Accessed on August 19 2022).

[20]    A. Kamilaris, F. Gao, F.X. Prenafeta-Boldu, and M.I. Ali, "Agri-IoT: A semantic framework for internet of things-enabled smart farming applications", *2016 IEEE 3rd World Forum on Internet of Things (WF-IoT)* Reston, VA, USA, pp. 442-447, 2016.
[http://dx.doi.org/10.1109/WF-IoT.2016.7845467]

[21]    R. Shrestha, G.F. Davenport, R. Bruskiewich, and E. Arnaud, "Development of crop ontology for sharing crop phenotypic information", In: *Drought phenotyping in crops: From theory to practice. Plant phenotyping methodology*, 2011.

[22]    R. Shrestha, R. Mauleon, R. Simon, J. Balaji, S. Channelière, A. Alercia, M. Senger, K. Manansala, T. Metz, G. Davenport, and R. Bruskiewich, "Development of gcp ontology for sharing crop

information", *Nature Precedings,* pp. 1-1, 2009.

[23]    C. Roussey, J.P. Chanet, V. Cellier, and F. Amarger, "Agronomic taxon", *Proceedings of the 2nd International Workshop on Open Data* pp. 1-4, 2013.

[24]    CTA, *Agrovoc multilingual agricultural thesaurus.* ICT Update, 2005.

[25]    A.C. Liang, and M. Sini, "Mapping AGROVOC and the chinese agricultural thesaurus: Definitions, tools, procedures", *New Rev. Hypermedia Multimed.,* vol. 12, no. 1, pp. 51-62, 2006. [http://dx.doi.org/10.1080/13614560600774396]

[26]    National Agricultural Library, "The national agricultural library's agricultural thesaurus", Available From:https:// agclass.nal.usda.gov/agt/ (Accessed on August 19 2022).

[27]    A. Soren, B. Christian, K. Georgi, L. Jens, C. Richard, and I. Zachary, "Dbpedia: A nucleus for a web of open data", *Semant. Web,* pp. 722-735, 2007.

# SUBJECT INDEX

www.ingramcontent.com/pod-product-compliance
Lightning Source LLC
Chambersburg PA
CBHW050838220326
41598CB00006B/390